AFTER
MEAT

AFTER MEAT

THE CASE FOR AN AMAZING, MEAT-FREE WORLD

KARTHIK SEKAR, PHD

All illustrations, figures, and tables were created by Karthik Sekar. Pictures of golden retrievers are reproduced via the FreeImages.com license agreement from photographers Hervé de Brabandère and Rachel Kirk. The Citric Acid Cycle figure was modified from one originally illustrated by WikiUsers Narayanese, WikiUserPedia, YassineMrabet, and TotoBaggins and is acknowledged under the Creative Commons license BY-SA 3.0. The Mississippi River figure was modified from one originally illustrated by WikiUser Shannon 1 and is acknowledged under the Creative Commons license BY-SA 4.0. Image of Thích Quảng Đức by Malcolm Browne for the Associated Press and available in public domain.

ISBN: 978-0-578-97737-9

Library of Congress Control Number: 2021913190

Front cover image by Julia Allum.

Book design by Tom Morgan, Blue Design (www.bluedes.com)

Printed in the USA

First printing edition 2021.

Published by Karthik Sekar

San Francisco, CA USA

www.aftermeatbook.com

Table of Contents

PART THREE: REPLACING ANIMAL PRODUCTS IN FOOD

PART FOUR: CATALYZING A FUTURE WITHOUT ANIMAL PRODUCTS

Introduction

Welcome to *After Meat*. I'm humbled and gratified to have your time and attention. I intend to make the most of it. I'm Karthik, a scientist in the alternative food space with a research career in biochemical engineering and quantitative/systems biology.

I suspect that you're roughly familiar with the moral and environmental arguments for moving away from animal products, or at the very least, you hold an inchoate sense that animal products are detrimental to the environment and are unethical. Perhaps you're already convinced, as I am, that the eventual replacement of all animal technology—man's use of animals as anything other than pets—i.e. as a source of food, clothing, medicines, or cosmetics—is inevitable. Specifically, you and I might believe that, in the future, humanity will completely eschew traditional animal technology and embrace the clean meat and clean protein revolutions, in which consumer goods are sourced from plants and grown via advanced cellular technology.

The movement away from animal-based foods already has tremendous momentum. Corporate fast food giants Burger King and McDonald's have both introduced veggie burgers sourced from the well-known, next-generation vegan food companies Beyond Meat and Impossible Foods.[1] The publication *The Economist* declared 2019 to be the "Year of the Vegan," claiming that a quarter of Americans between the ages of twenty-five and thirty-four declare themselves to be vegetarian or vegan.[2] The UK supermarket giant Sainsbury's predicts that a quarter of *all* Brits will be vegan or vegetarian by 2025.

All this being said, I suspect that you might be unfamiliar with the technological reasons for moving away from animal products, and that's the focus of this book. Simply put: raising animals for consumption is an awful technology. All indications suggest that the future of food will ultimately be tastier, healthier, cheaper, kinder, and better for the environment. This will happen *because* we won't use animal products.

In Chapters 1 and 2, we'll discuss a model for how technical progress works. In Chapter 3, we look at the cow as an example, and examine this animal as a bioreactor that society uses to produce steaks, leather, and milk. A cow takes more than a year to grow, and we "waste" more than ninety percent of what we feed the animal to reach the commercially desired body mass, due to the fundamental physics of cow biology (Chapter 4). These are irretrievably terrible metrics. We can do much better with alternative technologies, such as microbial fermentation, which will also be easier to innovate for process efficiency, taste, nutrition, and any other qualities we need or care about (Chapter 5).

An optional, more technical chapter (Appendix A) explains that perfect futurology is impossible, per the laws of physics, so I recommend reading the appendix before you start Chapter 5, which references the appendix. Even though we can't predict the future precisely, we *can* predict with confidence that traditional animal technology will be replaced based on the best technology and knowledge innovation model. Animal products *will* be replaced. The question is not *if*; it is *when* and *how*.

Further chapters explore issues adjacent or complementary to the technological argument. Chapter 6 explains that animal technology is not necessary for complete nutrition. Chapter 7 discusses how humanity derives and can recalibrate pleasure, an attribute often ascribed to meat-eating. Chapter 8 explains our culinary history and projects a resplendent future gastronomy that's free of animal products. Chapter 9 discusses the role of large institutions, such as government, in facilitating our innovation machine and specific plays for animal technology replacement. In Chapter 10, I explain what every person can do to move humanity beyond our reliance on animal products.

While this book focuses on the technological argument, I wrote it ultimately out of my compassion for animals and their welfare. I hope to accelerate the adoption of a similar attitude in the greater population. I suspect that, once technology has allowed alternatives to outcompete animal products and/or replaced them, we'll societally militate against animal products just as we did against child labor and lead paint. Eventually, we'll universally reevaluate how we treat animals, viewing our past behavior with despair and regret, similar to how we now chide ourselves for our history of tolerating slavery and subjugating women. However, I did not want the morality argument to distract from the main argument that a world after meat is an inevitable reality based on technological innovations. I have, accordingly, allowed this argument to take center stage in Chapter 11, the final chapter.

After Meat covers many topics: biology, physics, chemistry, philosophy, economics, policy, neuroscience, and engineering. The breadth and depth of these topics are supported with a summary and list of defined terminology located at the end of each chapter; though, you might want to read them *before* tackling the chapter. I suspect this will ease understanding, especially for the more technical first half of the book.

Despite my use of an absolutist tone and verbiage, I do welcome well-intentioned disagreement and clear refutations of anything presented here. Being wrong is part of the knowledge-generation process (Chapter 2), which I hold in the utmost regard. In fact, I take it a step further—I predict many of these ideas *will* be proved wrong or become outdated. Knowledge is ephemeral, much of it waiting to be replaced by more precise understanding as civilization and scientific exploration develops. Probabilistically, it's more likely for something in this book to be wrong simply because there are so many claims here. Therefore, please reach out to me via blog, email, or otherwise to engage about these ideas.[3]

Back and forth emails with beta readers have been some of the most fruitful and joyful parts of writing this book. If contacted, I cannot guarantee that I'll always respond, but know that I'll always appreciate your effort, especially if we both value knowledge generation in good faith: we seek the truth and the best ideas, and we are willing to change our own beliefs.

I will assert many controversial points, and I expect pushback. I hope some of the ideas will gain wider acceptance in the coming years, so much so that they'll seem unremarkable in twenty or thirty years. I ask for a nuanced approach to *After Meat*. Sure, even I might come to regard the chapter on nutrition as baloney within two months of release, but that doesn't automatically discredit the chapter on the intractability of animal technology. Most of the claims can stand independently. An anodyne, risk-adverse book will not push progress and solve as many problems. Instead, I've strived to make many interesting claims that can be proven false because ultimately that's how we all learn.

I hope you enjoy reading, and I would love to hear your thoughts.

Sincerely,
Karthik Sekar
June 1, 2021
San Mateo, California

PART ONE

Problems and Technological Innovation

CHAPTER ONE

Solving Problems

A BETTER CHEESE

In February 2017, I journeyed to sunny Barcelona, Spain with my mom and sister. We planned one long day to hit all of the major attractions: a walking tour downtown to learn about the city and its history; the monumental Sagrada Familia; topped off with an evening stroll along the beach. The beach excursion turned out to be longer and farther away than we anticipated. By the time we finished, it was late, and we were hungry. When I checked my phone for nearby restaurants, a vegan tapas restaurant popped up, and my sister, whose interest was piqued, implored us to try it. The individual dishes were all geared toward reproducing well-known meat-based tapas and Spanish food but in vegan form, including a paella with vegan shrimp, grilled potatoes, and vegetable skewers. The tapas were delicious except for the vegan cheese. Instead of accentuating the dish, it intruded. With thoughtless bravado, I commented that I could develop a better vegan cheese. My sister quipped, "So why don't you?"

I was stunned into silence; I had no good answer. Why didn't I? At that point, I was working as a postdoctoral researcher in Zurich, Switzerland. My earlier decision to pursue a doctorate in biochemical engineering proved consequential. I had all the requisite education and training necessary to attempt a viable vegan cheese–development strategy. After returning to Zurich, I spelunked through the scientific literature and the ongoing work in the vegan cheese space.

My first question at this juncture was: Can cow cheese be made without the cow? After all, cheese, like all physical things, is constructed of molecules that in turn, and in combinations, make up individual ingredients. Perhaps it would be possible to source these ingredients outside of cow milk. Everyone knows that cheese is mostly fat and protein, and these building blocks are everywhere—in plants, fungi, and microbes. In particular, a cheese-specific protein found in cow's milk—casein—imparts the unique, splendid properties of cheese: the stretchiness as cheese melts and the curdling of milk into a solid. If we could source casein from a microorganism such as bacteria, then we would no longer need the cow to make cheese.

Producing specialized proteins is a modern alchemy and a key thrust of biochemical engineering. The best-selling protein-based drug last year, Humira, a product of biochemical engineering, costs more than $100,000 for a single gram.[4] In 2017, there were 3 million prescriptions for Humira, which is used in the treatment of autoimmune diseases such as rheumatoid arthritis and Crohn's disease.[5] In contrast, a gram of gold sells for a mere fifty bucks. The production of a protein drug like Humira involves importing specific DNA from animals that encodes the protein of interest into a host, or suspension, that contains producer cells. These host cells incorporate, transcribe, and translate the DNA sequence to create many copies of the final protein. In short, the producer cells, once implanted with the specialized DNA fragment, become factories that generate the chosen protein.

This process is called **heterologous expression**, and it arguably started with the mass production of insulin at Eli Lilly in the eighties.[6] Up until that point, insulin used to treat human diabetes was distilled from the pancreas of a pig. In time, pioneering biochemical engineers learned that by using heterologous expression they could, instead, make bacteria produce insulin. To do this, they inserted the **gene** for insulin in the host bacteria. The gene, a contiguous sequence of DNA, contained all of the information necessary for the bacteria cell to produce insulin. The bacteria could then be grown in large bioreactors, the same vats in which we brew beer. For process efficiency reasons that I'll discuss in Chapter 4, this method produces the same insulin as the animal-based procedure, only more cheaply and cleanly, and without pigs having to sacrifice their pancreases.

In my own process, I pondered whether or not casein could be produced the same way that we produce insulin, Humira, and numerous other proteins. It turned out that this was not a new idea. There were already a few—albeit small—casein endeavors involving similar concepts. I also learned that the problem wasn't easy. Casein has post-translational modifications, meaning that, after the protein is synthesized by the cow, it's further modified chemically to confer additional properties. For casein, these modifications are critical in the assembly of protein molecules into a scaffold of larger spheres, or casein micelles.[7] Dairy fat preferentially resides within these micelle spheres, and dairy products could not exist without them. Butter is created by churning, where the micelles are physically broken apart, and the fat floats to the top. Cheese is created by applying acid or heat to denature the casein protein at the surface of the micelles. This denaturation induces the micelles to glob together and form the cheese curd.

The capacity to perform post-translational modifications depends on machinery unique to different organisms. Production microorganisms such as bacteria and yeast have limited capability for post-translational modifications compared to the metabolic process of a cow, and that capability is difficult to engineer into less-complex organisms, such as aforementioned bacteria and yeast. Producing Humira involves similar challenges, whose high cost is partially driven by production difficulties as well as the need to achieve sufficient purity for pharmaceutical use. In contrast, beer is easier and faster to produce, and does not need to be as pure; therefore, its costs are low and economical. Unfortunately, it seemed that the creation of casein was more like Humira in terms of cellular production, but valued as cheaply as beer due to the availability of cow's milk.

So I brainstormed alternate strategies. If cow casein is so hard to produce in microorganisms, why not try to find an already existing alternate casein? There are near-infinite numbers of proteins in the natural world. Surely, at least one of them could provide the functionality of cow casein? If humanity could measure the "caseinness" of different proteins, then we could screen a vast number in order to find the best possibilities. The highest scoring proteins would be our prime candidates in a non-cow–based production.

I consequently obsessed over ways to evaluate proteins in such a way, and quickly too. If it took an entire day to examine 100 proteins, then I was unlikely to find suitable hits in sufficient time. I ultimately settled on a high-throughput (more hits per unit of time) strategy that involved pooling a group of proteins together in order to see which ones formed into the macrostructures characteristic of casein micelles. The strategy would be similar to testing donated blood, which is only done in large batches in order to save time and money. If there's a hit, the entire batch is flagged. In the same way, I could test between ten thousand to 100 thousand proteins per day with the batching strategy. But there was a problem—how would I fund such an expensive venture? I calculated that just the instrumentation alone would require tens of thousands of dollars.

A few months later, an opportunity for funding appeared in my inbox with a subject line "Have an idea that could change tomorrow?" Certainly, I thought. The email further read:

> *The fellowship is designed for postdocs at home in sciences, engineering and social sciences who are willing to engage in a dialogue on relevant social, cultural, political or economic issues across the frontiers of their particular discipline.*

The monetary award offered was also substantially more than a typical fellowship for scientists at my level and sufficient to pursue the project. I could not have asked for a more fitting opportunity, and I set to work, building the best proposal I could.

As I developed the proposal, ideas and insights deluged my fevered brain. The idea of ending animal agriculture seemed so obvious, so inevitable, especially to someone with a biochemical engineering background like me. Preliminary calculations suggested that animals were an awful food-production technology, and too few scientists were attempting ventures to eventually replace them. Writing the introduction exhilarated me as I made my case, one scientist to many others: here is the current reality, here is where we could be, and here is a proposed means to get there. I was also spitting out the ideas to friends and colleagues, who seemed intrigued. I also felt that I had plenty more to say about it.

Despite my efforts, I was not chosen for the grant. Perhaps my publication record or resumé was insufficient. Maybe the proposal was too controversial. So, after the postdoctoral position, I briefly worked at a vegan food startup that seemingly offered the direct opportunity to tackle many of the challenges I lamented in the proposal. For various reasons, that venture also did not pan out. Afterward, I came back to circulating topics and writing this book about the technological argument for food production to move away from animals. This book has been percolating within me for years, and writing these ideas out is the necessary release.

NATURALISM AU NATUREL

I also write this book because I feel my views are not adequately represented by the animal rights movements. Vegans and vegetarians are often conflated with "all-natural" advocates, and for good reason. If someone is vegan, they're also more likely to skip chemical deodorant and discredit vaccines. They are more likely to buy something only if it's organic or all-natural. I dislike this. I'm unfairly lumped in with these groups when I wholly disagree with many of these ideas. For example, the "organic" label, while an effective marketing ploy, is ultimately a meaningless descriptor of food. The term itself does not say how much healthier or environmentally friendlier the food is because the term only pertains to how the food was made versus its nutritional content.[8]

The descriptor "natural" irks me the most. Naturalism is the idea that the more "natural" something is, the inherently better it is. For example, dish soap is better when sourced from natural ingredients. Monogamy is "unnatural" so we should have polyamorous relationships. And we should all run barefoot because that's what our forebearers did. Naturalism is constantly invoked with regard to ingredients in different diets, i.e. meat.[9] In fact, Naturalism is one of the four Ns used to justify eating animals: natural, necessary, nice, and normal.[10]

First problem with naturalism: what exactly is natural? We don't have an official definition of the term from the United States Department of Agriculture, which does formally define "organic."[11] The paleo diet has garnered popularity because it is presumed to be the natural diet of our ancient predecessors.[12] But if you eat a paleo diet because it's more natural, then which variety? Not all hunter-gatherer diets in every region had access to the same foods. Tomatoes and potatoes were only available in the Americas. Bananas were likely only found in Southeast Asia or possibly Africa. Furthermore, have you seen an uncultivated, ancestral banana? If not, check out **Figure 1**. What we find in grocery stores today is the outgrowth of generations of hybridization and selective breeding. Does a banana lose all that "naturalness" after years of human meddling?

Secondly, we only selectively apply naturalism when relative in a negative context. We don't apply it to modern medicine and reject heart transplants, knee replacements, or chemical drugs as undesirable because they're "unnatural." We don't waltz into a forest and declare that, because everything is natural, we can consume it all, because we know some of the plants are toxic. We luxuriate in air-conditioned homes, scrub away dirt, and shriek at the sight of vermin. In fact, our best science about disease and illness has been brought to bear to identify and conquer "natural" pathogens such as bacteria or viruses that make us sick. Washing our hands with commercial soap reduces the probability of transmitting infections from pathogens in food contaminated via our hands. These actions hardly strike me as natural. Our comfortable, fashionable clothing starkly contrasts with the trappings of our ancestors. Many of us sit double-digit hours each day transfixed by glowing rectangular screens. Conveniently, we do not apply the naturalistic ideal to these situations. In fact, in many instances, it has been far more beneficial for our species to eschew naturalism, as in the hand-washing example above.

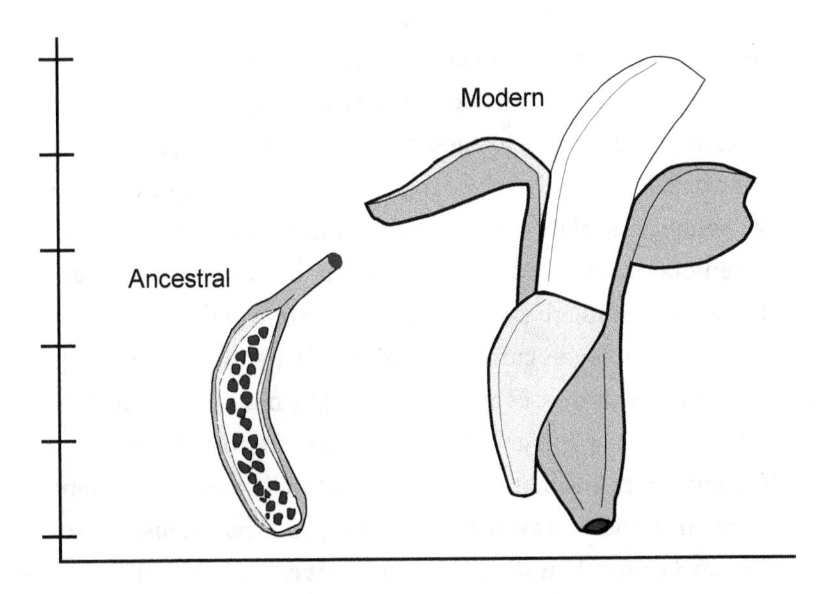

Figure 1. "Natural" banana versus the modern variety. An ancestral banana (left) juxtaposed with a modern banana (right) in relative scale. The modern banana was selectively bred over many generations to become larger and to have imperceptible, comestible seeds.

Instead, in the cases where naturalism seems to engender positive outcomes, we apply more precise principles rather than relying on the same to explain the benefits. For example, a paleo diet may indeed be healthier than a diet of just fast food. However, it's not the naturalness of the paleo diet that necessarily confers nutritional superiority. Less free sugars or lower glycemic indices in a paleo diet will better explain the diet's effectiveness than the perceived naturalism of the food itself. Similarly, barefoot running may indeed be better for us than running with sneakers because of how human biomechanics work: one study showed that barefoot runners experience less impact on their feet because of how a bare foot strikes the ground.[13] We do not need to apply the naturalism principle to market these concepts in such instances because more precise explanations—better knowledge—are available.

Clinging to the naturalism fallacy results in terrible consequences. The anti-vaccine movement routinely appeals to naturalism to promulgate its message. The subtext is that vaccines are not natural and are thereby dangerous, despite their obvious and visible benefit to society over the years. Homophobic movements often denigrate homosexuality by claiming that same-sex couples are unnatural.[14] Similarly, we're constraining one of the most promising technologies currently available, genetic engineering, due to retrograde impulses against genetically modified organisms (GMOs). (Chapter 5 will greatly expand on and refute the anti-GMO sentiment.) Ultimately, invoking naturalism as an argument will slow the transition away from animal products because it's so terribly imprecise when superior arguments exist. "Natural" is an empty, uninformative adjective used to market consumer goods, but often has little relevancy when examined in depth.

THE EVOLUTIONARY IMPERATIVE

Some groups and individuals subscribe to "evolutionary naturalism," arguing that evolutionary forces have shaped us into the beings we are today. These forces formed our minds, bodies, emotions, and values, for better or worse. In their opinions, our lives and actions should be about slaking these forces or, at the least, these forces should excuse certain behaviors. For example, eating animals is often seen as necessary for positive human evolutionary development. One theory suggests that human society evolved as a result of the effort required to hunt large animals and share the spoils.[15] This history occurred for most of the 200 thousand years since *Homo sapiens* first speciated, i.e. split off from *Homo erectus,* to form an independent species. Therefore, according to some advocates of meat-eating, as an established core facet of our evolutionary heritage, carnivore-centric diets are necessary to ensure our continued progress as a species. Again, this is a terrible argument. By that logic, it would be evolutionarily better to remain segregated in small tribes similar to those we populated for most of human history and forego the organized, vast global societies we have built. But if we did that, we'd lose the interconnectedness to tackle the enormous problems facing all of society, such as mass vaccination and climate change. In other words, we cannot always look back to go forward. The Theory of Evolution alone cannot dictate the life that we are to lead. However, the science behind the Theory of Evolution can explain how biological life changes over time.

Specifically, the Theory of Evolution explains how a species changes given the **variation** between members of the species and the **selection pressure** applied to them. For all biological entities, at least one selection pressure is obvious: the imperative to reproduce. If a species does not easily or eagerly reproduce, then that species dies out. Consequently, we have been shaped by evolutionary forces to pursue and enjoy sex. Obviously, if we had not been programmed in this manner, then our kind would have quickly died out. Likewise, we must eat to survive. Therefore, evolutionary forces shaped us to feel weakness and pain when we need food and energized and happier once we've satisfied that hunger.

How does evolution shape a species? Charles Darwin, the progenitor of the Theory of Evolution, has a famous example. Darwin formed his ideas about evolution by observing ground finches in the Galapagos Islands. The finches feasted upon seeds, and their different beak sizes led to separate advantages and disadvantages when eating different types of seeds. A large beak allowed the finch to break through hardy seeds, while the smaller beaks could more speedily devour small seeds. If an island had more small seeds, small-beaked finches would eventually dominate the area, eating seeds faster than their large-beaked brethren. Conversely, if the island's flora favored larger, hard seeds, the large-beaked birds would proliferate instead. The Theory of Evolution posits that, given variation among beak size in the finch species, a selection occurs. The birds whose beaks allow them to consume more food, more efficiently, will reproduce faster than the birds with smaller beaks. This variation is key. If there is no variation among beak size and all the birds had the exact same beak, then selection cannot occur, and the population will maintain an unchanging beak size across generations. The beak size variation, as we found out later, stems from variation in the DNA. Natural chemical mutations and biological inheritance change the DNA between finches, resulting in different beak lengths.

Suppose a consequential scenario: a volcanic eruption suddenly wipes away the small seeds, leaving only large, hard seeds. The change is permanent, and the seeds remain large and hard for years to come. Suppose that based on an average beak size of 1 centimeter, that beak sizes range from 0.8 to 1.2 centimeters before the eruption. Often such properties are visualized with a distribution (**Figure 2**) where most finch beak sizes occur at the mean, but a variation occurs around this value.

After the volcanic eruption, every finch is disadvantaged initially. All of them struggle to eat the remaining large seeds; however, not everyone struggles equally. The few with larger beaks at 1.2 centimeters sate their hunger slightly more easily and consequently are able to sire more progeny. In contrast, the 0.8-centimeter-beaked finches struggle more and are unable to sire offspring as quickly as even the average finch. As a result, in the next generation, the distribution of finch beak sizes shifts. This shift continues into the next generation and the generation after that until a new equilibrium is reached.

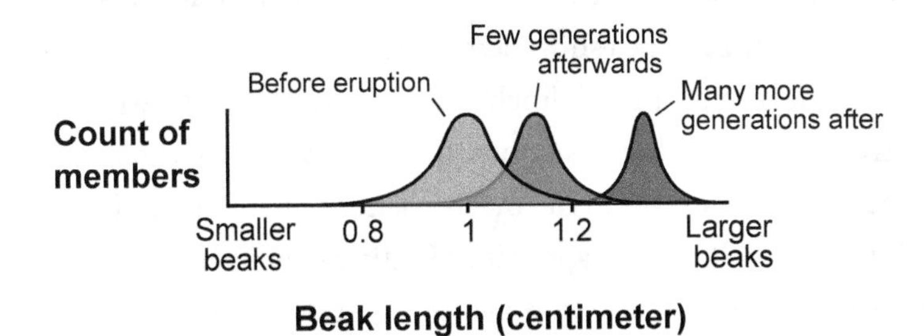

Figure 2. The shift of beak lengths after a volcano eruption. The leftmost curve highlights a distribution of beak lengths for the finch species. Most finches have a beak length of 1 centimeter, and few have 1.2 or 0.8 centimeters. After the volcanic eruption, the distribution of beak lengths shifts rightward (gets bigger) over the generations.

Obviously, there can be a point where too long a beak is disadvantageous, imaginably when the beak is so heavy it hampers flying or the finch's ability to move its head.

Another potential scenario is that the finches learn how to eat something other than seeds. Maybe they start eating cactus fruit. Initially, they're ill-suited, but once again there is some sort of variation in their ability to eat such fruit, such as harder skin to shield against thorns or sharper talons to rip apart the fruit. This trait once again potentiates over the generations, and after a number of generations, a new cadre of finches thrives with sharp talons, tough skin, and a predilection for cactus fruit. This new finch population enters a different **niche** of their ecosystem. Before they were in the seed-eating niche, and now they occupy the cactus fruit–eating niche.

To clarify how the Theory of Evolution is often misunderstood: there is no magical evolutionary "force" that shapes all members of the species equally. After the volcanic eruption that leaves only large, hard seeds, evolutionary forces do not magically lengthen the beaks of all the finches. There is no Jedi or deity that magically conjures this consequence into existence. The distribution of the beak sizes for the finches shifts across generations and the beak gets longer because the long-bill finches survive and reproduce more effectively than the short-bill ones, thereby passing on their DNA.

Because we understand the mechanism of The Theory of Evolution so well, we can apply it to finding new technology for future generations. In **directed evolution**, scientists find new proteins with desired functionality. I can apply a selection pressure to a pool of varied proteins to enrich the opportunity for proteins with a specific behavior, such as the ability to form fibers that texturize vegan steaks of the future. Over time, just as with the finch population, the protein population is enriched with proteins that survive the selection pressure—a measurement of some sort, perhaps, of how well proteins bind together into fibers or those fibers having the correct amount of strength. The selected proteins perform a function not found without that pressure in the natural world. In fact, I used such a directed-evolution strategy in the aforementioned alternative casein proposal.

So, it would seem that we have a lot to thank evolutionary forces for; without them, we would not be who we are today, both individually and societally. However, this does not mean we should regard our evolutionary imperatives with unquestionable reverence. In fact, we already refute some evolutionary imperatives for our ancestral niche. Consider reproduction, mentioned briefly above. Reproduction has been an evolutionary selection pressure for virtually all life at some point in time. However, this is not true if you live in any developed society today, especially if you're a fertile woman. Very few women in developed societies with access to family planning seek to be continuously pregnant or to mother as many kids as humanly possible. The opposite is likely truer, as constantly birthing and raising children takes a tremendous physical and economic toll on the health and wellbeing of a mother, and thereby her birthed children. This fact is affirmed in empirical data as well. As societies develop, people earn higher incomes and have access to water, food, education, birth control, and economic opportunities, the growth rate of the populations slows. Now, developed countries such as Denmark and Japan have a negative growth rate, meaning that the average woman is bearing fewer than 2.1 children.[16] This means that the population of these countries is diminishing. It requires, on average, 2.1 children per woman to replace the population. Even in emerging countries such as Bangladesh, the average woman now has 2.1 kids, down from a prolific 6.95 in 1970.[17]

The second selection pressure—starvation and food security—is being solved as well. Never before in human history have we had such access to abundant, rich foods. As Yuval Noah Harari highlights in *Sapiens*,[18] we're more likely to die from problems related to overeating (e.g. heart disease) than from starvation itself. In fact, to Harari's point, twenty-five percent of deaths in the United States stem from heart disease.[19] I concede that starvation is not a problem we've completely eliminated; there are still places, namely in Africa, where widespread famine persists. However, we have reason to be hopeful, as the trends point to an overall increase in food security and a corresponding decrease in widespread starvation.

We are solving (e.g. decreasing starvation, overcoming natural predators, surviving disease and infections) or rejecting (e.g. uncontrolled reproduction, tribal mentality) the selection pressures long imposed on our species by the Theory of Evolution. The Theory of Evolution cannot tell us what to do or not to do. Where do we go from here? Humanity has rationalized other problems to solve, and among those problems, it is my contention that overreliance on animal products merits our attention.

SOLVING OTHER PROBLEMS

After solving our own problems of starvation and predators, we are attempting to tackle other problems unrelated to natural evolution; for instance, extreme poverty. Jeffrey Sachs in *The End of Poverty* extols our pursuit and highlights the heartening progress in the reduction of poverty.[20] The book was published fifteen years ago, so I'll showcase the more recent, impressive numbers: According to the World Bank, the percentage of people living in extreme poverty has declined from 42.1% of the world population in 1980 to about 10% in 2015.[21] For China alone, 850 million people have risen out of extreme poverty in the same time span.[22] I fully concede that we have not solved all of poverty, for there is still around ten percent of people globally who suffer, but the overall trend is hopeful.

Our efforts to allay poverty and suffering have largely been intentional. Humanity purposefully eliminated smallpox using a combination of vaccination and coordinated medical efforts led by organizations such as the Pan American Health Organization, the World Health Assembly, and the World Health Organization. At the onset of those activities in 1959, smallpox killed more than 2 million people per year.[23] Approximately twenty years later, the disease was declared eradicated by the World Health Organization. The last known case was in 1975.[24] By eliminating smallpox, we've improved one of the many determinants for lifting others out of poverty, thereby promoting knowledge-transfer across generations. When folks are wiped out by disease, their knowledge is lost to subsequent generations. For example: an experienced farmer struck down by disease before she is able to convey her savoir-faire completely to her children. Her lost knowledge is tantamount to lost wealth and will only exacerbate her children's losses and/or poverty. Eliminating disease has not been the only accomplishment from organized global efforts to help eliminate suffering. With pest-resistant crops, cleanliness education, access to clean water, healthcare, and availability of mosquito nets, extreme poverty and its associated ills continue to wane.

However, we should not be satisfied only with helping people in extreme poverty. Sachs defined extreme poverty as the state where individuals did not know if they'd survive until the next day, such as not knowing their next food source. The next rung up the human development ladder is still poverty, just not as extreme. In this rung, food and water security spans a few days. Ideally, we'd continue to raise the standard of living for people on this rung so that they never worry about dying of hunger or thirst. In addition to tackling poverty, we are still working to raise other standards, such as clean, safe, and secure living spaces as well as basic access to health care and education. Nonetheless, we should feel emboldened by the progress of humanity. These problems are tractable, and directed and vetted efforts have paid dividends.

The subtext to *The End of Poverty* is that problems always must be solved in steps, starting with the most pressing need, which is generally water or food insecurity. Sometimes this requires a solution with obvious problems of its own. A business may set up a textile factory where workers spend long hours, pejoratively called sweatshops. Sweatshops are often derided for imposing terrible working and living conditions on workers. However, when critiquing sweatshops, we must always ask what the alternative might be. If the loss of the sweatshop means the people remain destitute and starving to death, then that suggests that the sweatshop, as bad as it is, solves a more direct and lethal problem, and is therefore part of net progress. The goal, of course, is to eventually elevate working conditions for all people. The next rung of the ladder.

In another example of this step-by-step view, consider the needs of an impoverished farmer in Brazil. Even though people in wealthier countries would prefer it, this farmer cannot prioritize rainforest preservation if he is struggling to feed his family or pay for medicine. His survival may be better served by burning the rainforest down to open more fallow land so he can grow more crops to feed his family. In this case, improving and securing his living standards would also help solve the environmental problem of deforestation. Given this kind of security, the farmer would likely not prefer the forest to be aflame, but without it, his hand is forced by the circumstances. Philosopher Will MacAskill estimates that a donation of a mere $105 to the organization Cool Earth,[25] which directly supports the education and income development of locals who inhabit rainforest areas, would offset the carbon dioxide produced in one year by an American adult.[26] Making the lives of rainforest inhabitants better ultimately solves the downstream environmental problem of deforesting the region.

Therefore, as we choose problems to tackle, we should be selective: we want to tackle the most pressing, tractable problems first in order to have the largest impact. I'm certainly not the first to raise this notion; in fact, the Effective Altruism movement has adopted and spread this philosophy since the late 2000s. We cannot market environmentalism and sustainability to destitute farmers in Brazil. We cannot instantly expect United States-level workplace standards for an area experiencing widespread, extreme poverty. We should, however, choose important, impactful problems to tackle. The inefficiency of animal agriculture is one of those problems.

How to supplant animal agriculture is not just an environmental or a moral problem, it's also an economic and technological problem. Replacing or superseding animal agriculture with non-animal resources can help alleviate extreme poverty as well as other "luxury" problems such as environmental issues. As we know, water and food insecurity are hallmarks of extreme poverty. Reducing water usage elsewhere would assuage such insecurities. Animal agriculture requires a tremendous amount of water, especially in proportion to the food it renders. A technology that provides the same food using less water would help curtail water inequities and extreme poverty.

I predict and assert that animals *will* be replaced, especially as protein products in the human diet. While I'm confident in this prediction, I cannot state how—exact futurology is impossible, argued in Appendix A. The fundamental physics of our universe simply makes such prognostication impossible in many areas, especially in regard to technological development. But we can proclaim how terrible certain technology is, especially when replacements are just visible over the horizon. In the next chapter, we'll discuss a future free of animal agriculture and the features of the technology that will make it possible.

CHAPTER TERMS

- **heterologous expression**: porting a DNA sequence from one organism to another
- **gene**: the minimal, contiguous DNA sequence needed to encode a protein
- **naturalism**: the fallacious, imprecise idea that there is something inherently good about what's "natural"
- **variation**: the spread of a specific trait, or the width of the distribution. For example, if human height was equally distributed between four feet and seven feet, then that would be more variation than if all humans were 5.5 feet tall
- **selection pressure**: any environmental feature that causes more reproductive success for a portion of a species based on trait variation
- **evolution**: an explanation for how a species changes given a selection pressure and variation within a species
- **niche**: the role occupied by a species, including what the species typically eats
- **directed evolution**: intentionally using the Theory of Evolution to impose a selection on a population (often a pool of proteins) to enrich for a subpopulation with a specific trait

CHAPTER SUMMARY

Since the modern Enlightenment (around the 17th century), increased knowledge and scientific discovery has enabled humanity to begin to solve the most exigent of our societal problems that threaten the natural evolution of our species, namely mass starvation, disease, and poverty. Fewer people today die or suffer from starvation, disease, or lack of water than ever before. We've also largely moved away from our evolutionary imperatives. Generally speaking, we do not seek to reproduce as much as possible and, in developed and emerging countries, are not satisfied with mere survival. We also eschew what's "natural," which has no precise meaning to us anymore despite the ubiquitous marketing.

At this point in time, innovators and global thinkers are following new pursuits, lifting the remaining 10% of the population that exists in extreme poverty to livable conditions, reducing hunger and disease in affected populations, and educating and training people for the work of the future. Problems must be tackled strategically and in stages. For example, trying to teach someone to read is fruitless if they are hungry and ill. Therefore, the most egregious problems must be addressed first. The world's reliance on animal agriculture is one of these fundamental problems, and we've already started endeavors to eliminate it. Supplanting animal agriculture with more sustainable plant-based protein creation portends cascading effects that will solve all sorts of problems ranging from water scarcity to all kinds of other environmental issues.

CHAPTER TWO

Knowledge and Technological Innovation

HOW DO WE KNOW SOMETHING?

At a previous job interview, I was presenting research from my postdoctoral tenure to three scientists. The atmosphere was heightened and adversarial, I suspect by design. Every claim I asserted was thoroughly scrutinized. I presented a method that detected bacteria undergoing division, and a deluge of challenges and questions immediately followed: "How do you know your measurement corresponded to the number of bacteria?" "How did you develop the method?" "How do you know this particular method corresponds to another more validated method?" I fully and directly answered the incisive queries of my data and methodology; however, a skeptical miasma still suffused the room. In the penultimate slide, I presented an explanatory mechanism that tied all of the data together and predicted when bacteria divided. To substantiate the mechanism, I presented two follow-up experiments to change the division timing, derived from the model presented. One of the attendant scientists rebuked me, asking: "Are two experiments enough to prove this model?" I verbally invoked a few more experiments that we tested and passed against our model. He then reiterated, "Are five to six experiments enough?"

How do we prove something? This question has occupied humanity's thinking for most of our recorded history. Aristotle differentiates true versus false in his treatise entitled *Metaphysics*, dating back from around 330 BCE. Stated another way, how do we generate knowledge? In the more formal terminology, we're asking how scientific **epistemology** works. We need new knowledge to make technological advancement; in the case at hand, to move beyond animal products. Research and development is how we'll fully move on.

Ask an ostensibly full-time knowledge generator/promulgator (e.g., a scientist or an academic) this question, and I sincerely doubt they could confidently provide an answer as evinced in the aforementioned story. In my education, the most that I learned about the knowledge-generation process came when I was in elementary school. We discussed the scientific method incessantly:

1. Question or problem: What are we trying to answer? What is the problem that we are trying to solve? We build from previous knowledge here.
2. Hypothesis: This is some conjecture that we make about the problem or question. It is something to be tested with an experiment.
3. Experiment: We perform some procedure or measurement used to prove or disprove the hypothesis.
4. Analysis: Here, we examine our data to see if our hypothesis was proven or disproven.

These instructions were thoroughly inculcated in me, particularly as manifested in the three different years I implemented them for school science fairs, so much so that I still remember them twenty years later. I remember designing my graduate research projects based on this same model, where I articulated the question, hypothesis, etc. Only within the last few years did I learn the formal name for this system of inquiry—inductivism. **Inductivism** implies that there are foundational knowledges, absolute truths, which we can build on and add to. The questions and observations must be anchored to underlying, validated knowledge. I imagine a brick house where scientists, or knowledge-generators, make the bricks. The bricks are then inspected by other reviewing scientists to ensure congruence. Very often, these reviewing scientists will return a brick to the originating scientists to refine it more. Only after the brick has been refined enough will it be added to the structure. The reviewing scientists consider each brick carefully; no spurious bricks can be added lest the entire knowledge structure buckle.

The most widely accepted knowledge model is seemingly inductivism, especially in the biological sciences. If it's not obvious, I allegorize the brick house to scientific publishing. And we might therefore automatically attribute perfect function to inductivism, if it's the underpinning to scientific publishing. But consider the following points:

If a scientist seeks to publish a study in a top journal, he or she first must entice a fickle editorial staff. If the study isn't within the scope of a currently hot topic or authored by a well-connected scientist, then the submitting scientist is rolling the dice. If the editorial staff accepts the study for publication, they'll send the submission out for peer review and solicit other scientists' views in order to gauge its impact and scientific rigor. There are, however, no defined criteria for scientific rigor, leaving such judgments to the whim of every reviewing scientist.

So, if the submitting scientists are lucky, the reviewing scientists will have no conflicts of interest and want to believe in the conclusions of the study. A friend once joked that publishing in a top journal requires a study that asserts something that the reviewing community has always believed in but for which has just lacked the evidence. Even if all of those parameters are satisfied, it is not uncommon for the editor and reviewers to ask for additional experiments and analysis, which sometimes may take years.

Defense of the current publication system often rests on a couple of fallacious arguments. One refrain is that pre-publication review makes a paper better.[27] Well, sure. If it didn't do *at least* that, then we would have zero reasons to maintain the process. The real question is whether it justifies the opportunity cost. For instance, one of my friends presented a project to me back in 2015. At that point, the key findings of the project were profound and backed up by strong experimental evidence. For career reasons, this colleague needed this work to be published in a high-impact journal. He spent years toiling away, polishing it more and more. He submitted it many times to top journals, getting rebuffed in the first year. After revising and timing his submission so that editors would be more receptive, the final submission—to the lofty journal *Nature*—finally gained some traction. Still, he had to spend an additional two years adding marginal experiments to satisfy the picky reviewers. Finally, in 2020, a full *five* years after he first showed me the project, the paper was published. The published version is certainly improved compared to the initial one that he shared, but is it five years, hundreds of hours, hundreds of thousands of dollars in added cost better? No.

Many problems with scientific publishing are underscored by this story. First, my friend spent an inordinate amount of resources to publish a paper that became, say, only ten to fifteen percent better. And the key findings were roughly the same. He could have instead been pursuing another relevant project, but due to professional incentives, needed a high-impact paper. The second issue about this story is, wouldn't humanity benefit from accessing these findings *sooner*? If a "less impactful" form of the paper had been released back in 2015, other scientists could have built on it, learned from it, and used it to advance their own science. This is especially important when the findings may affect something such as human health. If that's the case, then releasing findings in 2015 versus 2020 could actually save more lives.

Another friend mentioned a paper she wrote in graduate school, which still hadn't been published five years after submission. Her graduate school advisor, the person sponsoring the research behind the paper, is still trying to add more data to entice a top journal. The findings of this study relate to a protein behind the pathology of Alzheimer's disease. Isn't it tragic that humanity does not have access to this knowledge because of the demanding publication process and the career incentives? Talk to any publishing scientist and I'm sure you'll find stories like these. And unfortunately, we don't have a systematic way to find these "rotting in the drawer" projects. Proponents for journals may nonetheless argue that we need to make sure that the science is correct before release, but it begs the title question of this section.

So, there are clearly many deficiencies with scientific publishing as far as it is rooted in inductivism. At best, scientific publishing is a suboptimal way to produce knowledge or, in this case, get us to animal product alternatives. Additionally, the brick house analogy fails anyway because papers are often corrected post-publication, or even retracted. Furthermore, inductivism still doesn't answer the question: how do we know something? But maybe we can consider pieces of it, in particular, the aspect of scientists reviewing other science.

I've lost probably months of time in my research career due to inadequate quality control from scientists whose work I am building on. For example, I often worked with DNA sequences established in scientific literature. In more than one instance, I received DNA from other scientists to introduce into my bacteria for the purposes of eliciting specific functions (**phenotypes**). The scientists would also provide the text sequence of the DNA. The text sequence would explicitly designate whether or not the supplied DNA could impart the desired phenotype. I would introduce the supplied DNA into my bacteria, getting the bacteria to incorporate the outside DNA and impart its putative function. Often, I would not see the expected phenotype. My first inclination was always that I was doing something wrong with the outside DNA. I would spend weeks to months tinkering to ensure the proper conditions for the phenotype. Eventually, I would sequence the DNA myself and find that the sequence provided by the scientist was actually wrong. If I had simply sequenced the DNA first, I would have been spared all the failed efforts. Imagine that I'm trying to produce a non-animal steak from an antecedent scientist's work, and I get their DNA for the protein fibers. If I don't get a steak nearly as juicy as theirs, then I have to surmise that the DNA is perhaps aberrant. Confidence in the DNA sequence is essential.

As a result of such bad experiences, I changed how I sourced DNA from outside parties. I would first sequence the DNA myself if it were coming directly from another scientist, or I would rely on organizations that perform their own validation, such as the nonprofit AddGene. Within the AddGene website, I can see the DNA that both the submitting scientist and AddGene have generated. As a test, I bought some DNA through AddGene, and amazingly, it worked exactly as intended on the first test that I performed. Therefore, some sort of review process is indeed helpful and can accelerate our knowledge generation. And as AddGene demonstrates, new technology such as the internet can advance this. "Papers" are not the only unit of knowledge; that's clearly a vestige of pre-internet days. Our ability to find a new casein protein does not have to rely on scientific publishing.

THE EVOLUTIONARY MODEL OF KNOWLEDGE

There are also other problems with inductivism. The first is that we can never prove that something is absolutely true, no matter how many experiments we perform. Consider the case of boiling water. Suppose I wish to assert that water always boils at 212 °F or 100 °C. I could perform this experiment a billion times, and it would always be true. But if I ascend a mountain, this result no longer holds, as the water boils at a lower temperature. So immediately, the assertion that water boils at 100 °C is rejected by going up the mountain.

Karl Popper, an influential scientific epistemological thinker, recognized this fundamental asymmetry between proving and disproving assertions.[28] We can never actually prove something to be true, but we can *disprove* knowledge readily. Popper also recognized that all knowledge has limits and is provisional. The boiling water example has a clear limit with the elevation, or more precisely with pressure. When we change elevation, then we change the boiling point of water because the environmental pressure changes.

Any **knowledge** that we generate may be supplanted by better knowledge later on. With boiling water, we now have phase diagrams that tie the phase (vapor, liquid, or solid) to the temperature *and* pressure.[29] Another good example of the progress of knowledge is Newton's Laws. Anyone who has taken high-school physics has learned concepts such as force equals mass times acceleration, and that physical bodies will confer equal and opposite reactions. However, unless you've taken enough physics courses, you may not be privy to that fact that Newton's Laws have been replaced by superior knowledge—specifically, quantum theory and Einstein's General and Special Relativity. These theories are much more precise and have much more reach compared to Newton's Laws. For example, relativity will correctly predict a deflection that occurs as bodies orbit around each other; whereas Newton's Laws do not predict such phenomena. Newton's Laws are a special case of relativity, if anything.

All of our conclusions are ultimately temporary, a potentially depressing concept for knowledge generators and promulgators. Two comments to ameliorate such concerns: first, it is likely any generated knowledge will be a stepping stone to better knowledge, which can even fully replace that interim thinking. The superseding knowledge benefited from the pre-existing knowledge. That is something to be proud of. Secondly, knowledge can *still* be useful even when supplanted by better theories. We still learn and readily apply Newton's Laws because they provide useful calculations and inform engineering strategies. If I want to build a large bioreactor to produce my non-animal meat, Newton's Laws allow me to calculate how much power is needed to stir the tank. We understand Newton's Laws to be a special case of relativity, and we still learn them in school as an entry point into physics. We have not dispensed with Newton's Laws completely.

Popper's model for knowledge generation correlates well to the Theory of Evolution (**Figure 3**). We first develop knowledge through a variety of **conjectures** (akin to genetic variance) intended to solve a current set of problems. These conjectures are putative and have not been subject to a selection pressure. We reject knowledge that is unfit by applying **refutations**. The refutations can be arguments that point out the fallacies of the asserted knowledge or can be incongruent experimental results. Refutations *are* the selection pressure that I discussed in the previous chapter when examining the Theory of Evolution. As we saw earlier, all knowledge has limits. For example, the Theory of Evolution has been one of the sturdiest pieces of knowledge in the biological domain for over 150 years. It explains how, under given selection pressures, biological lifeforms will evolve. Evolution, however, cannot explain how life began in the first place. As of this writing, the inception of life remains a problem/question that vexes scientists. Accordingly, the inception of life remains in the new set of problems after discovery of the Theory of Evolution (in this case, Problems_{n+1}).

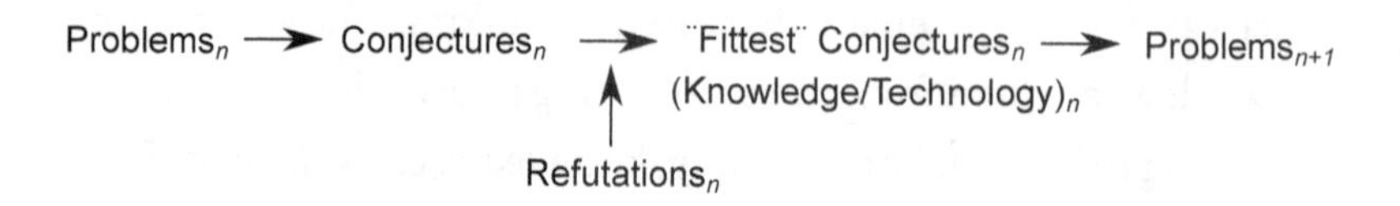

Figure 3. Karl Popper's model for knowledge/technology generation. The existence of problems inspires conjectures to solve the problems. Through refutations, the fittest conjectures are left. This constitutes a new set of knowledge and technology. With each iteration, we are left with a different set of problems to solve.

In the Theory of Evolution, fitness advantage is determined by whether an organism is fitter than the next best competitor. There is no "absolute" fitness. We cannot claim that elephants are more fit than sea slugs. Elephants would thrive more in dry climates eating leaves and fruit. Sea slugs thrive in aquatic environments eating algae. Both species occupy completely different niches and accordingly cannot be compared. Similarly, no current knowledge is absolute in the Popperian model. Therefore, knowledge *cannot be judged absolutely*. There is no known bedrock truth—there is nothing to anchor what we know. We can only evaluate a piece of knowledge in *relation to the next best competing knowledge* in how it solves a given problem.

For example, the Theory of Evolution's next best competitor may be Lamarckism. For the uninitiated, Lamarckism claims that biological traits are passed to descendants by the performed actions of the parent. Lamarckism supposes that ancestral giraffes had to stretch their necks in order to reach the leaves in the trees. The stretched neck was passed down to descendants, and each consecutive generation stretched more and more. Lamarckism does not require the variance that the Theory of Evolution demands, just a continued action of progenitors that will exaggerate the traits, eventually to the most useful degree.

Popper's model suggests that we are dramatically replacing our knowledge all of the time. In actuality, instead of wholly replacing a theory, it's often better to amend it, leaving the revised theory better than the competitors.

Here's what this can look like in the real world. Recent epigenetic research suggests that there may be something such as a Lamarckian effect within Holocaust survivors.[30] Specifically, there are chemical modification (epigenetic) changes to the DNA of Holocaust survivors such that they exhibit more stress stemming from the concentration camp conditions. These phenotypes were then passed down to the survivors' descendants. The descendants exhibited more stress compared to control groups, something akin to a Lamarckian effect. An uncorrected version of the Theory of Evolution would not predict such an effect. But rather than refuting the Theory of Evolution completely, we can add the "but." The Theory of Evolution still explains the selection of organisms when variance and a pressure exist, *but* via epigenetics, a Lamarckian effect can occur where phenotypes are passed that stemmed from the experiences of the parent. This amended Theory of Evolution has more reach than the original theory and is still better than any known competitor (e.g., pure Lamarckism) for the same niche.

Now, suppose we wish to maximize knowledge generation, such as finding ways to produce steaks from yeast. Hopefully, I've substantiated that inductivism is inferior to the Popperian model for knowledge generation. Therefore, we should leverage the Popperian model to propose a better system for scientific inquiry: any scientist or individual should be incentivized to make conjectures. In effect, conjectures generate the phenotypic variance for evolution. Likewise, we should incentivize scientists to refute any knowledge, leaving the "fittest" knowledge, so that we constantly iterate and generate knowledge faster and more transparently.

All of these activities should be public. For example, I should be able to access a website to see the conjectured knowledge and the responding critiques without an impeding paywall. It could be an explanation for how casein proteins form into micelles with the help of surrounding salts. The conjecturing scientists could then continue to refine or substantiate their conjectures with experimental data, repudiate the critiques (e.g., did you consider this mechanism?), and the next iteration proceeds. In fact, the mathematics and physics communities have seemingly embraced such a model on platforms such as ArXiv.[31] ArXiv allows any scientist to upload and publicize papers without any peer review. Many influential papers never go to official peer review; they simply remain on the ArXiv website. Other scientists can post refutations, and the original scientists can post amendments.

Thankfully, the chemistry, medicine, and biology research communities are seemingly heading in the same direction with the advent of the analogs ChemRxiv[32], MedRxiv,[33] and BioRxiv.[34] We've seen MedRxiv and BioRxiv take off during the 2020 coronavirus pandemic, as crucial knowledge needed to be circulated quickly.[35] Some scientists have raised concern over the rush, and the sloppiness of the studies, evident in the number of retractions.[36] But this is a step in the right direction. We ultimately want to be able to conjecture and reject science quickly, especially with a pressing problem in a coronavirus pandemic where the number of lives saved grows with available knowledge developed in the previous week.

In order for a Popperian system to become the norm, we need to develop the other half: refutation. Currently, scientists mostly do refutations through anonymized, blind peer-review, and this is insufficient. There is no way to publicly applaud their efforts nor critique them if it's in some way a poor or unfair review. Secondly, this system leads to bad incentives with the possibility of refutation devolving into a cartel-like system of mutual admiration: if you review my paper and give me an easy pass, I'll do the same for you. Furthermore, do I want to inflame another scientist who may review my grant for funding or my abstract submission for presentation at a conference? No. I'll have more professional success by being a Pollyanna than doing what's best for generating knowledge.

A BETTER PICTURE OF KNOWLEDGE

Instead of a brick house, better metaphors for knowledge might be as follows: a measuring implement such as ruler; a wayfinder (like a compass); or an elemental analyzer that can quantify the protein content of a food. In the case of a wooden ruler, we can measure the length of another object. The ruler will always have a limit; it cannot measure the height of a mountain or the length of a single water molecule. There is also a certain degree of precision. Our ruler might be able to say if the object is closer to 64 centimeters or 63 centimeters, but it probably can't resolve the difference between 63.333332 and 63.333333 centimeters. Likewise, a compass and elemental analyzer will have analogous limits for their respective measurements. A new iteration may be something, such as a laser ruler, which has increased precision and may have additional functions such as being able to measure the level of a surface.

I note a few more ancillary, but informative, properties of knowledge. We've discussed the comparative property, that the Theory of Evolution explains the change of species in environments better than Lamarckism. We've also discussed the limit of a piece of knowledge, that the Theory of Evolution cannot explain the origin of life or the life we are to lead. Karl Popper highlights falsifiability. The more falsifiable a piece of knowledge is, the more likely it is truer and more useful because it has survived multiple encounters with refutations. David Deutsch highlights "hard-to-varyness" as a more precise way to describe falsifiability.[37]

Let's look at an example. Suppose a piece of knowledge asserts the existence of an Abrahamic God, and all phenomena occur due to the will of this God. At face value, it seems to be an impressive piece of knowledge. In terms of reach, it can conceivably explain everything. But the outcome of God's will is easy to vary. It can adapt to any situation. Suppose we ask why a loved one dies in a car crash; one can say that was because of God's will. Suppose the person survives the car crash? We also can say it was because of God's will. The "God's will" assertion can be varied easily to fit any situation. Harking back to the ruler analogy, it's as if the ruler is made out of a soft clay. We can place it next to a variety of objects—extruding or compressing it as necessary—to obtain a length measurement. However, we concede that any measured number has limited usefulness because the ruler was so easy to manipulate, i.e., vary; it is more reliable in terms of outcome to measure length using the sturdy wood ruler, even though it lacks the flexibility and stretchiness to measure curved objects. Likewise, in a car crash, invoking the will of God offers zero guidance to outcome. Instead, we'd rather know the context: what were the road conditions? Was the driver wearing a seatbelt? We know that wearing seatbelts means better chances of survival. Such a hypothesis has been regularly tested against data, and the hypothesis has not been refuted. We can trust it more as a piece of knowledge *because* it withstood falsification attempts. We cannot trust anything similar about a "God's will" explanation for an event because of the lack of such falsifiability, nor can we do anything with it because it's so easy to vary. I implore interested readers to explore Karl Popper and David Deutsch's oeuvres for more depth regarding the desirous properties of knowledge. Deutsch, in particular, highlights the supremacy of explanatory knowledge, which meets all the traits we seek.

How does technology relate to knowledge? *Technology is tautological to Popperian knowledge*; technology is knowledge that is directly intended to solve what we, as people and society, consider problems. Technology can be a vacuum cleaner that solves the problem of cleaning dust out of carpet more easily, or a novel steak that better meets our nutritional needs. Technology can also be software. For example, a computer algorithm may also be put to use predicting flavor or aspects of food texture. There is nothing physically tangible about such an algorithm, but it can help develop new foods. *For the rest of the book, I will use knowledge and technology interchangeably.*

Technology, as with any type of knowledge, can be refuted. The infamous, recently dissolved company, Theranos, shines (or tarnishes, perhaps) as a prime example.[38] The Theranos leadership touted their blood-testing equipment as having the capacity to measure vast amounts of data merely from pinprick quantities of blood. The promise was revolutionary and disruptive: users could measure all of their blood levels every day, rapidly, and for a fraction of the blood and cost. However, in reality, the technology never worked as boasted. Instead, data was generated from competing equipment using diluted blood drops. The Theranos leadership clearly hoped that their science would catch up to their façade. Despite receiving nearly a billion dollars in funding, they could not develop the promised technology and lied in order to prolong their pursuits. Here, the refutation system failed due both to duplicitous actions of the Theranos leadership and the Pollyannaish outlook of uninformed investors and business partners. The earlier we can determine a technology as unfit, the better so we can allocate our resources elsewhere.

There is also the fact that technology, in solving one problem, can also create others. I immediately think of social media, which has solved some problems of connectivity and keeping up to date with families and friends. But social media also creates the problem, in seeking profitability for investors, of trying to compete for the attention and the data of the user. Social media companies strive to subject users to as many targeted ads as possible. As a result, they seek to entice new users and to capture our attention for as long as possible. Videos chosen by algorithm will automatically play after the previous one finishes. Enticing posts are selected to go to the top of the feed, based on the data and activity of the individual's usage. Product managers also know that our quest for "likes" triggers our addictive dopamine response, thus keeping us glued to the platform.

However, we do not need to completely eschew social media altogether. This is a selection pressure problem. The current market milieu, consumers, and regulations incentivize the companies to pursue such destructive strategies. We need to somehow adjust this selection pressure so that social media companies solve more problems and cause fewer new ones. I will not talk about the problems that technologies generate. I acknowledge them, and I agree that we should do our best to minimize them. Provided we implement the right selection pressures such as anti-monopolist regulation and consumer demand, we should be able to find a balance where technologies provide benefits while reducing the total magnitude of related problems.

Just as with any other type of knowledge, all technology is provisional. Better technology replaces previous iterations because it solves more problems or satisfies the selection pressures better. When new technology breaks through the market and threatens to render current technologies obsolete, we term it **disruptive technology** (**Figure 4**). Disruptive technology proliferates saliently in the electronics space. Smartphones with touch screens, such as the iPhone, have largely displaced the older BlackBerry-style phones. Cars replaced horses. Tractors replaced oxen. Once enough technical advancement proceeded, Digital Video Discs (DVD) completely supplanted Video Home System (VHS) tape cassettes and rendered them obsolete. Now streaming services have rendered DVDs a superannuated technology in turn.

If all technology is provisional, there must be a reason for that. I assert that all technology has a ceiling. The **ceiling** is the best that a current technology can conceivably become at solving problems before another disruptive technology of an entirely new design or paradigm is needed. This is equivalent to the notion that all knowledge has a limit. Consider a horse-drawn carriage, used to transport people and goods for the majority of human history. In fact, before the advent of automotive and locomotive technology in the 19th century, humans and animals were the only technology to physically move goods on land—prime movers—as author Vaclav Smil labels them and highlights in much of his work.[39] Trains developed before automobiles, and still animal carriages were needed for local, last-mile transport. Clearly, animals as prime movers have limiting factors: they need to be fed, to sleep, and require training through coercion. Moreover, they are limited in how much horsepower they can provide. None of these limitations applied to trains and automobiles. Once those technologies reached sufficient fruition, the technology of animals as prime movers waned. We will discuss the limits of animals as a food technology in Chapter 4.

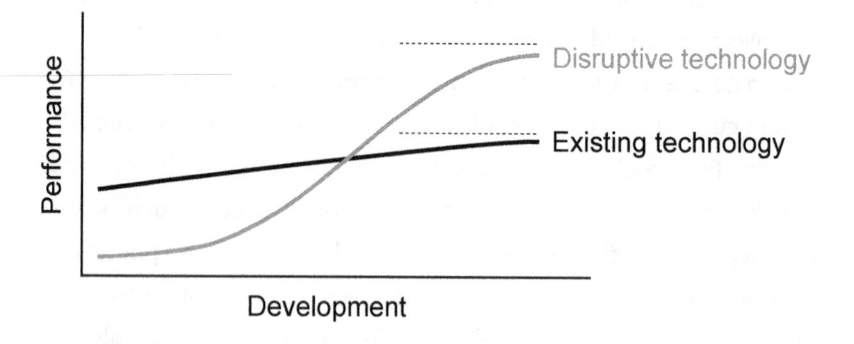

Figure 4. Disruptive technology and ceilings. Disruptive technology may not initially perform better than existing forms, but after enough development—iterations to improve upon it—the disruptive technology exceeds and replaces the existing one. All technology has limits, indicated here by the dotted lines. The ceiling for the disruptive technology must be higher than the one for the existing technology.

How do we generate new technologies that will disrupt old ones? Just like knowledge generation, this requires conjectures. I do not have a good theory for how this works comprehensively. If I did have an actionable theory, I would be a rich man. I offer instead a few thoughts. It's clear that humans have a particular ability to assimilate observations, process our surroundings, and make conjectures. According to the philosopher Daniel Dennett and his theory of consciousness of Multiple Drafts, our conscious perceptions fundamentally work by making conjectures about situations and applying evidence to refute them.[40] This is one reason why different people can have completely different perspectives on the same situation. This **consciousness** modus also explains hallucinations and mirages, situations that occur when the evidence fails to dispel the perception. Curiously, we are always employing the same knowledge generation process posed by Popper. We are natural innovators. A mechanistic biological understanding behind this process is still insufficient. Otherwise, we might be able to program an artificial intelligence that would be conjecturing and refuting, and presumably in a more facile manner than humans. For now, we humans are particularly capable of generating knowledge and technology as a direct outgrowth of our conscious makeup.

Additionally, all explanatory knowledge has **reach**, also posited by David Deutsch.[41] If we have knowledge that is explanatory, that is, if we can state the underlying elements, the specific interactions, and expected outcome, then that knowledge *reaches* to other domains. For example, with Newton's Laws, such as force equaling mass multiplied by acceleration, we could apply this knowledge to civil engineering in order to construct buildings. In fact, we can think of all engineering as simply extending the reach of other more "fundamental" knowledge, often in unexpected ways. The study of particle physics theory led to the development of the Positron Emission Tomography scanner to detect cancer. The best biological engineering tool, CRISPR-Cas9, was developed from the study of bacterial immune systems. It may be used to create all of our meat in the coming years.

In Appendix A, I discuss how the lack of predictability in future knowledge leads to the lack of predictability in our future generally. The property of reach plays into this. Sometimes the reach will not even be immediately apparent. Fundamental scientific techniques in nanotechnology, for example, did not receive much attention until nearly fifty years after initial publication.[42] The advent of new microscopy and imaging methods facilitated a resurgent interest in them. Generally, we cannot forecast how new knowledge ripples through our lives, and the property of reach only complicates this notion further.

Technological progress in certain areas seems to occur much more quickly than in other areas. I remember how rapidly computers seemed to improve in the late nineties and early aughts. I remember how anxious I was about buying a computer for high school. I knew that I would have to bite the bullet of buyer's remorse at some point. The laptop my parents bought me would be obsolete within a year. I could have saved all of the Best Buy mailing flyers and created a flip book showing a physical manifestation of Moore's Law—the number of transistors per area (density) doubling every two years. Compared to other areas such as biotechnology and nutrition, computer hardware innovation, particularly in data storage capacity and processor speed, demonstratively occurred faster.

I term the degree of difficulty of knowledge generation as **tractability**. A tractable problem is one that can be readily solved with modest knowledge generation. For example, suppose that I need to calculate how many apples I would have to buy in order to have one for each day of the week. Assuming that I go to the grocery store once per week, I can calculate that I need to buy seven apples during that excursion. The problem is so tractable that I don't even have to ponder it deeply; immediately, my brain conjures the number seven. Tractability of problems also directly relates to the existing knowledge/technology that we already have available. For example, suppose I need to measure the protein content of a certain food. Three hundred years ago, this would have been impossible even if we had understood what proteins were back then. Now, there are a bevy of instruments and biochemical assays one can perform. Today, even the most modest biochemistry lab will have the capability to measure protein content, however roughly.

Problems clearly vary in how tractable they are. We can take this a step further and consider the substrate itself and the problems surrounding it. Perhaps the most studied substrate in existence is the human body. We are clearly vested in the human body (a normal Joe might just call it obsessed with our own health) as exemplified by the resources we allocate to medical research—over $250 billion in 2012 worldwide.[43] We clearly seek to improve the functioning of the human body as much as possible: to make it live longer, work more efficiently, and, when systems fail, more easily fix them. However, if one compares a human body to an object like a car, differences emerge. We've been able to iterate cars so that they last longer, are less polluting, and are safer—at a fraction of the innovation cost (around $50-60 billion in 2012[44]) that we've spent on humans. It is safe to call cars a more tractable substrate than humans. How tractable a substrate is will suggest how readily we can iterate or improve it in the knowledge generation apparatus.

Lastly, we allocate a different amount of resources to different problems. Resources may take the form of people working on the problem, the money that we invest to do that work, or even the computers running simulations in order to generate more knowledge related to the problem. Collectively, I call this **bandwidth**: the resources that we apply to a problem in order to generate more knowledge. How we allocate bandwidth as a society depends on a mixture of current problems and values. For example, during World War II, the US had trouble accessing natural rubber due to the Japanese occupation throughout the Pacific Theater (where that resource was produced), specifically in Indonesia.[45] The United States made a concerted effort to discover synthetic sources of rubber and invested heavily in petrochemical research. As a result, and with partnership with Dow Chemicals, the synthetic rubber industry was born and rapidly burgeoned. In fact, today, we mostly use synthetic rubber in place of natural rubber. Also, during World War II, the US shifted bandwidth toward the development of nuclear arms. And hence, the atomic bomb was created. Both problems were clearly tractable but lacked the requisite bandwidth to solve the problems until they became prioritized.

I find that most laypeople with whom I discuss scientific development see bandwidth as the primary limitation to knowledge generation. That is, the only reason we haven't made substantial scientific developments in certain spaces is because we do not apply any or enough resources to these problem areas. I disagree with this sentiment. Yes, the tractability of the problems absolutely matters. We haven't cured cancer because it's a difficult, intractable problem. We pour tons of resources into this pursuit; the US National Institute of Health spends over $5 billion per year,[46] pharmaceutical companies invest almost ten times more.[47] Collectively, a substantial amount of bandwidth is spent on cancer research. While we've made progress in this area, it is trifling compared to the magnitude of resources that have been invested. Many researchers have concluded that we would achieve better returns elsewhere.[48] I do not disparage those who have put forth such a Herculean effort. I see it more that cancer is a particularly intractable problem owing to a multitude of factors such as the mechanistic differences between different cancers, the localization of drugs and treatment, and the detection and diagnosis of the disease.

In contrast to cancer research, I assert that there exists a subset of problems that are tractable with modest bandwidth. From the title of the book, you'll be able to glean one of those problem areas. Replacing animals should be a relatively tractable task, as I will discuss in Chapter 5, but we have allocated a pittance in terms of bandwidth (see Chapter 9); accordingly, we are still availing ourselves of inferior animal technology. The next section of the book delves into the details behind current animal technology and why it is so ripe for disruption. Animal technology has a low ceiling.

I have also delivered a summary of what I find to be the best model for knowledge/technology generation. This foundation for humanity's knowledge generation enterprise will constrain and inform the rest of the book. Because of the amount of preliminary information, I leave **Table 1** as reference to the aforementioned concepts. Note that the term "Pareto frontier" will be discussed in depth in Chapter 4.

Table 1. Knowledge in many forms.

TERM IN EPISTEMOLOGY	TERM IN TECHNOLOGY	TERM IN BIOLOGY
Knowledge	Technology	Species or library
Conjecture	Formulation of a technology	A variant of a species
Refutation	Government regulations, Customer demand	Selection pressure
Iteration	Iteration	Generation, round, or iteration
Limit	Ceiling	Pareto frontier
Problem	Problem or market (in economics sense)	Niche
Nucleation	Nucleation	Speciation
Tractability	Tractability	Variation and speciation capability
Bandwidth	Bandwidth	Number of entities within the species or library

DESCRIPTION	EXAMPLES
Anything that can be subjected to Popperian model: solves problems and can be refuted. Can also be iterated.	Theory of Evolution, Newton's Laws, a vacuum cleaner, a computer algorithm
An entity attempting to solve a problem or survive/reproduce in a niche	A new theory, Theory of Evolution, prototype vacuum cleaner, established vacuum cleaner
A force that favors a subpopulation of conjectures or variants based on some trait	An experimental measurement that favors one theory versus the others. A carbon tax.
Creating variation on a piece of knowledge and technology, then applying refutations, in order to try to solve more problems	Theory of Evolution with epigenetics. Better cars, better computers, better phones
Absolute coverage of problems a knowledge or technology can solve before needing replacement	Theory of Evolution can explain how species change over time but cannot explain the origin of life. Vacuum cleaners cannot find me a job.
Whatever issue or question we deem worth solving. Something that knowledge or technology can be applied toward.	Starvation, disease, dirty carpets, how do species change over time?
Incepting a new type of knowledge or technology to solve problems	Isaac Newton formulating calculus, Darwin formulating the Theory of Evolution, First prototype cars and phones
How difficult solving a set of problems is. How difficult it is to nucleate/iterate a given knowledge or technology.	Cancer is a much more intractable problem versus developing synthetic rubber. The human body is a much more intractable substrate compared to cars.
The amount of resources allocated to solve a problem	Academic research funding. Venture capital funding. Number of people working in a given area.

CHAPTER TERMS

- **epistemology**: the philosophy behind knowledge and what we know
- **phenotype**: a measurable or observable biological trait
- **conjectures**: a putative or durable knowledge
- **refutations**: data, experiment results, or an argument that can falsify a conjecture in comparison to a competing conjecture
- **knowledge/technology**: anything that solves problems or answers a question
- **ceiling**: the limit of a technology
- **bandwidth**: the amount of resources allocated to solving a problem
- **consciousness**: the act of repeatedly conjecturing a perception of a situation, falsified by encountered evidence
- **reach**: the property of explanatory technology to extend to unforeseen niches or problems
- **tractability**: how easy it is to develop knowledge or technology in a market, problem area, or around a substrate (e.g. human bodies or cars)
- **disruptive**: a new technology that displaces or can displace an existing technology
- **nucleate**: beginning a new technology

CHAPTER SUMMARY

The generation of knowledge/technology will determine how quickly we can replace animal products. First, we must understand how new knowledge/technology forms, and can leverage Karl Popper's evolutionary model, the best-known knowledge-generation model to date. Knowledge and technology are developed through the process of conjectures, positing a putative idea or invention. We try to refute the knowledge, and the most "fit" knowledge endures. We can nucleate new technology and iterate through conjectures/refutations until we reach the ceiling. Once the ceiling is reached, we must find technologies of entirely new designs for the specific problem/niche.

PART TWO

Animals As A (Terrible) Technology

CHAPTER THREE

Processes and Competitors to Animal Technology

THE PERFECT PROCESS

When I was in college, I obtained my first Apple computer, a first-generation MacBook Pro. From then on, I was firmly indoctrinated into the Apple universe. I loved the sleek, minimalist design, the sturdy construction, and intuitive, Unix-based operating system. Immaturely, I argued with friends and acquaintances about why Macs were better than their competition. And I naturally followed Apple-related news, especially regarding products. Rumors of an Apple smartphone had been spreading through the computing world, and finally, on January 9, 2007, during an Apple keynote address, Steve Jobs introduced the iPhone to universal approbation.

The iPhone reveal was a watershed moment. At the time, I was in college, and few peers had a smartphone. Those who did would often buy one without a data plan and use it to connect to the campus Wi-Fi. Pre-iPhone, smartphones were entirely of the BlackBerry mold where the screen would take perhaps half of the surface, and the other half would be the separated keyboard with plastic, recessed buttons. These phones were largely viewed by my peers as neat, but strictly functional—an instrument catering to bankers, financiers, and CEOs—and whose purpose was to respond to emails during the taxi ride from the airport to the client meeting. With previous smartphones, not too many of my friends pined for one, but that completely changed with the iPhone. I worked as a teaching assistant during the summer of 2007, and one of the students had an iPhone. I noticed that, when other students first were introduced to that student, the greeting was generally followed with a request to see and interact with the iPhone.

The rest of this story is well known. The iPhone's design was reproduced by Apple's competitors and rapidly antiquated the push button, BlackBerry style. Smartphones also became far more ubiquitous. In 2010, while working for AmeriCorps on my picayune salary, I got my first smartphone, the HTC Incredible. Within five years, the norm turned from not having a smartphone to having one. The smartphone is perhaps the most striking and obvious example of the technological achievement of today compared to twenty years ago. I recall a Reddit.com discussion thread where a prison lifer was released after decades and witnessed a smartphone for the first time in befuddled amazement.

The iPhone was a striking leap forward compared to competitors. Somehow Apple's approach to the fundamental concept of a smartphone was vastly more prescient and advanced than everyone else's. In the traditional paradigm, a design team first creates a prototype.[49] This prototype is then passed to the engineering team. The engineering team notices intractable features in the initial design, and then strips out those features. After engineering, manufacturing examines the now-stripped down design and finds other features incapable of implementation in the manufacturing process. The design is culled even further, leaving an inferior, distant simulacrum of the initial prototype. The product is only improved upon in the next design cycle. In contrast, Apple could iterate their designs much faster, thereby reaching a design friendly for manufacturing. Apple *designers* performed complete cycles of their design, in the process engineering prototypes all the way to trial manufacturing.[50] By testing engineering and manufacturing earlier, the constrained, iterated design was amenable for production. Apple's innovation wasn't just the perfect product; it was the perfect production process and the rapid path to get there. Other innovative companies such as Dyson follow a similar paradigm of many manufacturing-friendly iterations before the first sale.[51]

Intuitively, production and scalability should strike us as necessary precursors for a successful industry. Let's suppose that the iPhone was first created by designers only, in a process where they handcrafted the device. Obviously, that would never work in the market, especially for a smartphone. It would be far too labor intensive and costly for Apple to employ so many iPhone makers, not to mention the problems of quality assurance. Apple would never be able to circulate a handcrafted device to the masses. In the end, the iPhone's ubiquity also maximizes Apple's bottom line, as the second greatest fraction of their revenue is now App Store purchases, presumably much of which comes through iPhone usage. Without a process to produce smartphones at scale, the world would be a different place today.

CHEMICAL ENGINEERING

Entire industries will form and disappear based on the existence of a suitable process. For example, during the early European Industrial

Revolution, industrialists needed a more efficient way to make alkali, a common ingredient in soap. Up until that point, soap makers had burned seaweed and harvested the alkali from the ash. In 1789, a French scientist, Nicolas Le Blanc, developed a process to convert salts directly to alkali.[52] Once tax breaks on salt were offered, the salt-to-alkali revolution crested.[53] Today, we continue to make alkali directly from salt.[54]

Humanity rarely appreciates how important process is, probably because we hardly ever see or notice it; with a few exceptions like when we visit a production site by taking a brewery tour. At the start of the product's lifecycle is creation, when we require an efficient production process to chemically transform the raw inputs. As discussed though, transformation alone is not enough; the process must be efficient, scalable, and profitable. Enter the discipline of **chemical engineering**.

Traditional chemistry operates on a small scale. Perhaps you've taken a chemistry lab class in high school or college. You may remember a glass beaker or a cylindrical measuring flask. The same tools are used throughout a professional chemist's career. These are small-scale tools. A chemist may synthesize and/or characterize a particular compound in small batches, often an amount less than a serving of oatmeal. If a chemist figures out the synthesis stages to make an exciting compound, perhaps a new drug, it's a first step to bringing a product to consumer market. We cannot produce chemical compounds by using a comically large glass beaker: how would we mix the darned thing? How can we ensure that the temperature is evenly distributed throughout this vast beaker? We can't; we need a new design and science.

Naturally, a chemical engineer's education focuses on solutions for producing chemicals at a large scale. How can we ensure proper mixing? How can we separate a particular offending compound from our end product? The education includes courses such as **kinetics** (studying and calculating the *speed* of chemical reactions and phenomena), heat transfer (how fast temperature transfers through a medium), and separations (how to split mixtures into desired and undesired components). Instead of beakers and flasks, a chemical engineer will employ reactors and distillation columns. A typical college senior, as part of the curriculum, will design an entire process in order to produce a particular compound as their capstone project.

In particular, **reactors** are large vats that control the conditions necessary for the designed process. They are more scalable than a glass flask; specifically, parameters such as temperature and pressure can be controlled throughout much larger volumes. Reactors can reach multiple stories and hold over 100 thousand liters, roughly the volume of a double-decker bus. These reactors employ a large number of sensors and actuators to maintain the set conditions.

Let's say that we're trying to temper chocolate, forming bars from hot, melted chocolate. The liquid chocolate becomes solid as the cocoa butter within crystallizes upon cooling, though an assortment of different crystals are possible (six to be exact).[55] Two types of cocoa butter crystals result in soft, crumbly chocolate. Another two types melt too easily, and the sixth type is too hard. Chocolatiers strive for crystal type V, which imparts a glossy sheen, satisfying snap, and melts at body temperature—in your mouth.

Chocolatiers can preferentially form V crystals when the chocolate is cooled at 34 ° Celsius. If it's cooled at a lower temperature, the chocolate will be crumbly. A higher temperature makes the chocolate too hard. Furthermore, we want proper mixing so that the crystals form uniformly throughout. Therefore, we fit our reactor with an impeller that rotates and mixes the chocolate smoothly. To control the temperature, we turn on the heat jacket of the reactor to evenly apply heat, and a temperature sensor communicates when the mixture gets too hot or cold and adjusts accordingly.

Processes do not merely satisfy consumer demand, but often swing history as we discussed last chapter with synthetic rubber development. When you buy the latest LeBrons, you'll find that the sole is mostly synthetic rubber. On the other side of the war, the Axis (Germany) desperately sought oil in order to fuel their war machine. They conducted a folly-ridden invasion of Russia in order to secure oil reserves.[56] Operation Barbarossa was so disastrous that it arguably turned the tide of the war against the Nazis, especially since it gave the poorly prepared Russians time to regroup, rearm, and then counterattack.

As with LeBlanc's alkali innovation, sometimes an existing process has to wait for other conditions to change (e.g. salt taxes) before becoming viable, like the Haber ammonia-producing fertilizer process. The Haber process falls within the chemical engineering categorization, a process to rapidly make ammonia from hydrogen and nitrogen gas. Agriculture is a process to put food on our plates. For the longest period of our history, we struggled to produce food efficiently and reliably. As we found out later, it was difficult to supply a key ingredient—ammonia—to the plants. We relied heavily on animal technology, in the form of manure; however, this was still too inefficient and could not adequately meet our demand.

In the early 20th century, the Haber process was developed, named after the chemist Fritz Haber who discovered it. If hydrogen and nitrogen air molecules interact in our atmosphere, ammonia will only infrequently be produced. Scientists and engineers learned that the reaction occurs most efficiently under pressure, at a high temperature (500 ° Celsius), and with the help of a **catalyst**, specifically, iron powder laced with other trace compounds. The catalyst is not consumed in the reaction but instead increases the kinetics of the reaction, or the speed at which it occurs.

The hydrogen and nitrogen molecules will **adsorb** (stick) to the catalyst surface, allowing them to more easily rearrange and consequently form into ammonia. Fritz Haber demonstrated proof of concept of his process by building a pressurized reactor with catalyst in 1909. BASF, a German chemical company, bought the technology, and by 1910 the process was scaled up to meet industrial demand and eventually enabled the growing of the food that nourished and formed many of the people whom you know and love today. Once again, the product itself (ammonia) wasn't the determinant; we already had it with manure. Rather, we needed an efficient, scalable ammonia-production process to make it widely available.

In the last chapter, I discussed the idea of knowledge and technologies replacing older, inferior counterparts. Processes are a kind of technology and, accordingly, are subject to the same forces: The Haber process supplanted the competing but inefficient Birkeland–Eyde and Frank–Caro processes, which had replaced animal manure. Synthetic rubber displaced natural rubber as the consumer product standard of choice. Both processes are satisfying clear needs. The Haber process enabled a source of nitrogen for agriculture, and the synthetic rubber industry enabled an electrically and thermally insulating bulk material that is also chemical resistant, pliable, and durable.

In a congruent manner, we can view animals as a production technology. We have already replaced many animal products such as movement (prime movers), ammonia (manure), and fat (oil) for lighting applications. As author Paul Shapiro highlights in *Clean Meat*,[57] whale oil was used as fuel for lamps in the mid-19th century. At one point, the whaling industry was the fifth largest industry in the United States,[58] but the advent of petroleum-based fuels such as kerosene put the final nail in the coffin of the whale-oil industry, which had all but disappeared sixty years later. Nevertheless, animal products are still widespread today for gastronomical, nutritional, and some biomedical research needs.

ANIMAL TECHNOLOGY

Humans transitioned into settled societies starting approximately twenty thousand years ago,[59] where tribes maintained stationary homes and conducted hunter-gathering forays into the wild. However, sedentary populations depleted local resources rapidly, and a new source for food was required. About twelve thousand years later (or roughly 10,000 B.C.E.), plants were deliberately cultivated and maintained in order to feed a growing population. As far as meat goes, hunting still dominated as the primary food source.

Hunting, likewise, was an unsustainable process to meet the demand of the growing population. Hunters deplete the game on the land around them to the point where, because wild animals do not reproduce fast enough, the distance hunters have to travel becomes too far to reliably fill appetites. Naturally, these initial societies sought to domesticate animals and rear them for meat. The best candidates for domestication satisfied specific characteristics: animals that grew quickly, but with sufficient mass by adulthood (so not small birds).[60] The animals had to be tamable and social, thus easier to herd. Carnivorous species were a non-starter because they would require two kinds of husbandry, one for the prey and and one for the carnivores. Ruminant species such as cows could consume abundant grass, widely available at zero or low cost.[61] The omnivorous pig was able to eat anything and fatten up quickly, making it ideal to accompany Christopher Columbus on his foray into the Americas.[62] Likewise, chickens reproduced readily, could eat just about anything, and were easy to manage because of their docility and limited ability to fly.[63] They also provided a near-daily, secondary stream of food in unfertilized eggs.

The commonplace animal products of today would not have come into existence unless the animals themselves had been easy to bring to maturity with superlative process metrics including cheaper input, fast growth, and more meat for less effort. It's rare to find a pigeon filet or alligator stir fry other than as menu oddities. Cows, pigs, and chicken are the primary supply of animal products today *because* they lend themselves to better process-engineering compared to other species. We tried cultivating a variety of species, but ultimately these three satisfied our selection the best; the choice was not based on some innate preference.[64]

Your mind might be saying, "Well, I've tried pigeon meat, and it just isn't very good." This misses a key point: domestic animals today are the outgrowth of generations of selective breeding. Breeding equates to the directed evolution experiments I highlighted in Chapter 1. A cow will generally vary in a trait, or such variance can be introduced via crossbreeding with another species to produce a desired trait (like growth rate). Similar to the banana, the cows we consume today are vastly different from those found on our ancestors' dinner plates. Today, cows are even bred for the precise distribution of fat marbles in their flesh or how tender their meat is.[65]

For a long time after we transitioned to agrarian societies, meat was rare and consumed more rarely compared to the preceding hunter-gatherer days where diets could exceed 90 kg (200 lb.) of meat per year, depending on the region.[66] In Ancient Rome, poor individuals consumed an estimated 5 kg (11 lb.) of meat per year.[67] The situation was even more pronounced in China during the early 20th century, where poor farmers consumed on average 0.3 kg (0.7

lb.) per year, often during New Year's or wedding celebrations.[68] For reference, today most of the developed world consumes generally over 50 kg (110 lb.) per person annually. [69]

At the turn of the 20th century, a shift in animal technology occurred, as documented in Vaclav Smil's *Should We Eat Meat?*[70] The new Haber process, which allowed increased crop production, also resulted in surplus feed for animal agriculture. Similarly, the development of refrigeration technology, especially in ships, enabled cheaper meat export and import via sea-based trade. As a result, the industrialization of animal technology emerged, where consolidated slaughterhouses located near centralized railheads processed far more animals than a typical farm. Early slaughter-houses were notoriously unsanitary, famously documented by Upton Sinclair in *The Jungle* in 1905, eventually leading to the Meat Inspection Act and Pure Food Drug Act passed the next year. These acts raised animal treatment and hygiene standards.

Soon, with the industrial revolution hitting full swing, incomes generally rose, and demand for meat grew. Confined animal-feeding operations burgeoned, in which the animals are housed in climate-controlled structures that automate feeding, plumbing, and waste removal. As of today, such operations dominate animal agriculture worldwide except for in Africa and, in first world countries, on the very small percentage of farms practicing organic agriculture. As of 2010, sixty percent of all pork and seventy-five percent of chicken came from confined animal feed operations. These operations are process-wise more efficient compared to traditional animal agriculture (rearing animals on land) and require less manpower in order to produce more meat.[71]

Starting with a mature, live steer (we largely consume castrated male beef cattle and not females) that weighs in at 450 kg (1000 lb.), the animal is first killed, and the blood is drained. The serum from the blood can be separated from the red blood cells, and after more separation, specific proteins and other biomolecules can be extracted and sold. The protein heparin, for instance, is separated from the blood and sold as an anticoagulant for profits of billions per year.[72] The skin is often used for leather manufacturing. After draining the blood and skinning the corpse, the carcass will be about 270 kg (600 lb.). About 160 kg (350 lb.) of this will be the consumed meat or the animal muscle, a matrix of ordered protein more or less marbled with fat. That leaves 110 kg (250 lb.) of bones and additional fat. The structural protein collagen can be separated from the bone. Collagen can consequently be broken down into gelatin; both are used quite extensively in cosmetics and gelatinous foods (e.g. Jell-O). The fat, also known as tallow, was used for frying and making margarine, but that has fallen out of favor over health concerns in the nineties.[73]

While the animal product industry makes the most of money from meat sales, the **allied products** of collagen and the blood-based products ultimately add to its bottom line, too. For the pork industry this represents roughly six percent of the value and ten percent for beef.[74] This matters because the industry has thin margins: Tyson Foods disclosed an operating margin of 6.5% on its bovine industry in 2018.[75] For perspective, Apple[76] and Coca-Cola[77] boast over twenty percent profits for the last couple of years. This means that Tyson Foods has less leeway when problems happen, i.e. they're unable to sell as much meat or allied products. Realistically, Tyson would have to raise prices or find a different product to sell in such an event. Accordingly, if animal product producers could not sell the allied products to the cosmetic, research, or pharmaceutical industries, they would have to charge higher prices.

Ultimately though, all animal products are formed from a few classes of large biomolecules: mostly protein with some fat and other smaller molecules sprinkled in that we care about for nutrition (e.g. vitamin B_{12}) or for taste (e.g. heme). I've highlighted cases of specialized animal-based proteins, casein and heparin, which garner specific value.

Despite what the consumer may think, many animal products are needed just as bulk protein—the specific protein isn't important as long as it is protein. For example, the bovine serum albumin (BSA) protein can be separated from the serum component of cow blood. This protein is sold to researchers as a coating agent and standard; it helps quantify other proteins by mass. BSA is a consumer product because it's abundant (representing fifty percent of the serum's mass) and otherwise waste from the animal industry.[78] Its actual function is inconsequential other than being able to dissolve well in water. Similarly, ground and processed meat is leftover protein and fat mashed together. Again, the chemical details and functions of the protein do not matter in this context.

Food producers likewise do not have to be too concerned with the specific type of protein in some products, especially with the recent explosion of alternatives to animal products. The food company Beyond Meat takes proteins from pea plants to form plant-based sausages and burgers. The important feature of pea protein is its abundance and scalability compared to other plant protein sources. Chemically, pea protein is unexceptional to Beyond Meat other than being available and cheap. Impossible Foods pursues a similar course although it sources soy and potato protein instead. The company JUST released a mung bean-based alternative to egg omelets. The JUST egg is a liquid, but then solidifies on application of heat. Any dissolved protein will eventually **denature**, forming clumps that separate from a liquid as a solid mass, under the right conditions. Certainly, some proteins are more able than others, but JUST leveraged this basic biochemical fact to perform the egg transformation with mung bean protein. I worked in a startup where we accomplished the same result with blends of pumpkin seeds. You might worry about nutritional differences between different kinds of protein, and we will discuss further in Chapter 6.

Certainly, other animal-based products provide a specific function. Collagen provides a "gluey" property, and gelatin is known for its gel-like properties. Heme confers the slightly metallic taste that apparently is beloved in burgers. Impossible Foods sourced heme from a non-animal source to incorporate into its beef-like veggie burgers.[79] Cow and pig fat tend to be semi-solid; at room temperature, they are solid, perhaps even spreadable like butter. They liquefy immediately with application of heat, browning and crisping the meat around them. For Impossible Foods and Beyond Meat, coconut oil has proved the most available and capable substitute.

In summary, animal products can be thought of as providing mainly protein, both bulk and specialized, a semi-solid fat, and other molecules nutritionally or gastronomically of interest like vitamin B_{12}, zinc, and heme, respectively. Any technology that replaces animals will invariably provide these constituents more quickly, more cheaply, and at a higher quality. But of course, as we discussed in the last chapter, we must always pose an alternative. Without an alternative, nothing can supplant animal products. I see potential for replacement with an ancient technology—fermentation engineering.

FERMENTATION ENGINEERING

Some preceding examples of chemical processes (e.g. soap, rubber) capture only the drudging, lifeless processes of formal chemical engineering, but that's not always so. Earlier I remarked that brewery tours are one of the few examples where a production process is gleefully brought to the public view. Brewery tours have proliferated as craft (small production) beer exploded in popularity in recent years.[80] Cynically, I think brewery tours are an excuse to drink beer under the pretense of sophistication and supporting local business; nevertheless, I'm heartened at any means to showcase a chemical process, especially a biochemical one. Brewing is an incredibly old process, as often noted on such tours. Archaeological evidence suggests that the Ancient Egyptians brewed beer (4 to 5 thousand years ago).[81] The oldest operating brewery is Weihenstephaner in Germany, active since 1040.[82] Brewing processes continue to exist successfully into the modern age, and if anything, are even more widespread because of their outstanding metrics. Brewing falls under a class of general biochemical processes called fermentation engineering.

Once you complete one brewery tour, you've done them all. First, the grain (typically barley) is steeped in warm water to generate the wort, which is a liquid that contains sugars liberated from the grain during the mashing process. Solid particulate is separated from the wort, akin to running a mix through cheesecloth. The brewers then boil the wort and add hops, a pungent flower that imparts bitterness and helps pasteurize what will become the eventual beer, preserving it from spoilage for long duration. Finally, the mixture is cooled, and yeast is added. The yeast cells ferment the beer; in other words, they consume the sugar for their own nutrition.

Metabolizing sugar is a combustion reaction where energy is released, akin to burning fuel or wood. Combustion reactions release energy in the form of electrons, but the electrons must go somewhere; otherwise, combustion cannot occur. For most combustion, oxygen can accept the electrons (and turn into water in the process). That's why a bonfire is stoked when you fan in more oxygen. The yeast cells in the closed environment of a beer bottle lack the available, receptive oxygen in which to dump their electrons. Instead, biological life must release electrons into other molecules, such as ethanol, to complete the combustion reactions. Humans have this ability, too, though we make lactic acid instead of ethanol. In sum, the combination of available sugar and low oxygen means that the yeast produce copious amounts of ethanol, otherwise known as an alcoholic beverage, in the process.

Brewing takes advantage of a variety of reactors. Go on any brewery tour, and one set of the reactors is obvious, the large metal vats where wort is boiled, cooled, or fed the yeast. I remember touring the Miller Brewing Company in Milwaukee and seeing merely the tops of the large, metal reactors as they protruded through the floor from below. The less obvious reactors are the small ones, imperceptible to the naked eye—the yeast themselves. Yeast cells have control systems like a large-scale mechanical reactor, with microscopic fungi maintaining their internal pH and salt content in order to maximize their performance. Also, as often found in reactors, yeasts contain catalysts in the form of **enzymes**. Enzymes are specialized proteins that facilitate chemical transformation. In yeast's case, a collection of enzymes helps turn sugar into three essential ingredients: ethanol, which makes the beer into an alcoholic beverage; carbon dioxide (CO_2), which confers the delightful fizziness; and more yeast, which accelerates the brewery process. The reactor analogy is so apt that during my research career, I often deployed the same math used to assess large-scale chemical reactors to evaluate microorganisms such as yeast.[83]

Additionally, the brewers only need to add what seems a trifling amount of yeast, and the final beer will have much more yeast than at the start. Typically, the brewer will add about two to six grams[84] of dry yeast (same weight as a couple of ping pong balls) to a wort mixture of twenty liters (roughly five gallons). At the conclusion of the fermentation, which may take a few days to a few weeks, the number of yeast cells may replicate to number roughly twenty times what was originally added.[85] Most of the yeast is dead; however, the dormant cells can be revitalized and used to make another batch of beer. The yeasts can continually perpetuate themselves because of their **autocatalytic** property—the ability to reproduce oneself, i.e. yeast producing more yeast. In fact, this property enabled yeast to become a staple ingredient for many thousands of years since the start of the agricultural revolution.

Technologically, yeast is not just limited to help make beer. Yeast helps bread rise. Again, the baker just has to add a little bit of yeast starter to the dough. The yeast will consume the starches in the bread, spitting out the CO_2, thereby leavening the bread and making it rise. Knead some dough, then cover with a cloth to preserve moisture and put in a mildly warm place, then come back and it seems that there is roughly twice the dough compared to before. Yeast also can be added to mashed grapes, transforming the sugars into alcohol; however, wine is often fermented over longer periods at low pressure with the resulting carbon dioxide insufficient to make the liquid fizzy.

Yeast is not the only microbe used for fermentation. Cultured dairy products such as yogurt, kefir, and cheese require the addition of microbes to convert the milk sugar (lactose) into fermentation products, primarily lactic acid. Similarly, many other foods are the result of fermentation, including obvious ones such as sauerkraut, kimchi, and some tofu to non-obvious ones such as cured sausage, chocolate, coffee, and Tabasco sauce. Interestingly, fermentation food processes have not been replaced by other technology. They have been one of the most enduring food production processes since our ancestral years. While we've generally dispensed with old cooking styles such as cooking over an open fire, fermentation methods reign supreme, and there is still much more that we can reap—and learn—from this ancient technology.

To appreciate the power of fermentation-based autocatalysis, consider the correlate with money. A grandfather offers his grandson two choices for an inheritance payout:

• Choice 1: $100 the first year, $300 the second, $500 the third, and so forth for twenty-five years.

• Choice 2: $.01 one year, $.02 the next, and $.04 the third, and doubling every year after that for the same twenty-five years.

So which choice garners more money? Choice 1 provides $62,500 total, but Choice 2 provides nearly *five times* more at $335,544.31.

Formally, Choice 1 is **linear** growth, and Choice 2 is **exponential** (**Figure 5**). We calculate the change over time by considering each iteration. For Choice 1, each step entails addition to the previous value, specifically the new payout is $200 more than the year before. For Choice 2, the new value is *multiplied*. With each iteration, the previous value is multiplied by two; therefore, these values grow faster over time. Indeed, for Choice 2, most of the money comes from the latter years, with the 24th year paying out $83,886.08 and the 25th year $167,772.16. In the same way, the amplification with investing occurs too, albeit, far less dramatically. I'm not sure how many people have grandfathers who guarantee a hundred percent return every year. But consider a more modest, realistic example: if you invest $1,000 now into a SPY index fund, and it averages nine percent return per year, then in forty years it'll be worth about $36,500.

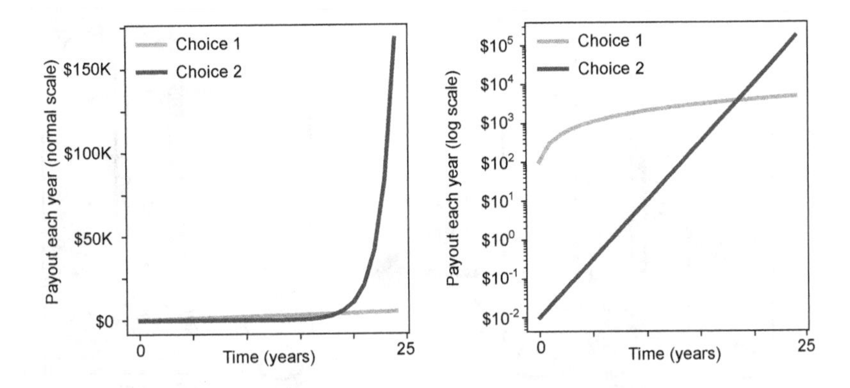

Figure 5. Linear versus exponential growth. Two profiles of Choice 1 and Choice 2 for the inheritance payout. On the left side, we plot the two curves on a typical (linear) axis. The right side plots the same data but uses a logarithmic scale where each tick increment is a multiplicative increase over the previous one. On log scale plot, exponential trends look linear.[86]

In the same way, yeast cells can multiply themselves with each iteration. Because most food conditions are hardly the ideal scenario for maximum yield, yeasts do not always double themselves (as anyone who's ever had their bread loaves come out looking like flat bricks can attest). Generally, yeast will double themselves faster in conditions replete with oxygen and good mixing, i.e., controlled bioreactor conditions (**Figure 6**). In such an environment, yeast can double themselves as quickly as every ninety minutes.[87] To highlight how powerful such speedy autocatalysis is, suppose we started with a single yeast cell, which is twenty picograms, about a quarter the mass of a single red blood cell. Then suppose we had a bioreactor large enough to cultivate the yeast and enough liquid medium to nourish the doubling yeast throughout indefinitely. With all that, we could have the equivalent of Earth's mass in yeast within just eight days.

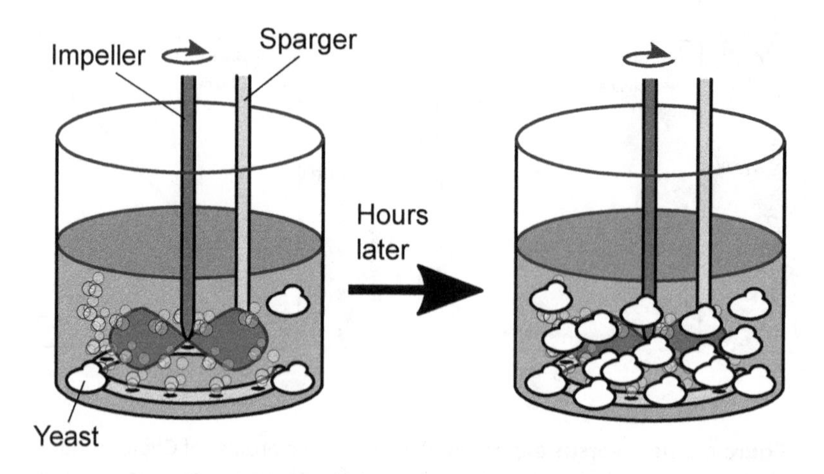

Figure 6. A bioreactor. A bioreactor controls the environment of microorganisms in order to optimize the environmental conditions: availability of nutrients (e.g. mixing by the spinning impeller), bubbling oxygen (e.g. through the sparger), and controlling pH/temperature. Typically, the reactor is seeded (inoculated) with some microbes, which then double rapidly.

It may be worrisome that yeast cells could potentially replicate themselves seemingly interminably. However, sugar will run out, and the yeast will become starved. Furthermore, as I stressed before, the conditions have to be ideal. Otherwise, the yeast may not grow fast enough, and the majority of the time, there's no growth at all. Yeast is not the only organism capable of such exquisite autocatalysis. Virtually any microorganism is capable too, including bacteria and other single-celled fungi. We have even been able to take cells from human kidneys, insects, and cancer and suspend them in reactors and cultivate them for many doublings. Some bacteria double themselves incredibly quickly. For example, *Escherichia coli* is the most well-studied organism on the planet, mainly because it is so easy to grow. In the optimal conditions, *E. coli* bacteria can double themselves every twenty minutes, meaning that a single cell can reach the mass of the Earth in just four days, compared to the eight days for the yeast under perfect bioreactor conditions.

Note: you've probably heard that *E. coli* is antagonistic to our health. There is some nuance here. Most *E. coli* strains are beneficial and abundant within our gut. In fact, there is a good chance that you have copious *E. coli* in your body right now. There are *E. coli* subspecies that produce Shiga toxin. Shiga toxin-producing *E. coli* (STEC) are worrisome, but generally they are the miniscule minority of all *E. coli* that are spread from fecal matter.

Given the metrics, microbes and fermentation processes should intrigue us as we consider alternatives for creating bulk and specialized protein. First, how much of the microbes' mass is protein? Quite a bit, it turns out. Microorganisms are approximately fifty percent protein by mass. So just in terms of bulk protein, they are more than up to the task. To produce specialized proteins, we can heterologously import desired genes into the host. As discussed earlier, we've done this for insulin, and actually have already started for heparin.[88] In fact, in both these cases, the microbial source is cleaner and cheaper than the animal-based method and has essentially supplanted the animal sources. Similarly, Impossible Foods imported the genes for heme catalysis into yeast; engineered the yeast organisms' metabolisms to produce heme; and consequently, the heme can be harvested from the same bioreactor processes used in beer production—and not from beef blood. Certainly, the microorganisms cannot do everything yet; however, there are companies—New Harvest, Clara Foods, Perfect Day, and Geltor to name a few—working specifically to produce casein, egg albumin, whey, and collagen/gelatin using microorganisms.

And what would a meat-producing fermentation process look like? We don't have too far to look: consider the single-cell protein, alternative meat product Quorn.[89] The bulk of Quorn is sourced from the fungal *F. venenatum* single cell organism, much like yeast, grown in bioreactors. After screening this fungus from a variety of candidates in the 1960s, Quorn was extensively tested for safety and toxicity for the next fifteen years. Toxicology tests were performed on eleven different animal species and a human trial with 2500 people.[90] Quorn is arguably the most safety-tested food product on the planet, yet it was not until 1985, after that extensive testing, that Quorn became available to the European public.[91]

In terms of process, the *F. venenatum* duplicate themselves in a bioreactor and then are heat treated to reduce their DNA and RNA content, per imposed safety criteria.[92] The mixture is then **centrifuged**, or placed in a spinning chamber to separate out large components from smaller ones. At this point, a fungal protein paste remains and can be shaped and set like any other industrial animal-meat paste. You might not be familiar with this principle, but it is the step before chicken, pork, and/or beef mash is formed into those recognizable nuggets, hot dogs, and patties.

There is room for improvement in the Quorn process. Egg albumin is sometimes added to the fungal paste to help solidify the end product. Instead, what if we could have two different organisms in the fermenter, one producing the protein and another producing the albumin? We might also have other strains for taste, structure, or to add health supplements. If we have GMO acceptance, then we can imagine more possibilities. The fungal cells could directly produce the albumin, taste molecules, or vitamins, if we change their DNA to do so.

Fermentation-based food production works superlatively because of the autocatalysis of the microbes. Certainly, animals are capable of autocatalysis. They do reproduce more of themselves. In fact, this reproduction is a huge process advantage. The producers merely have to ensure that the animals are fed and kept well enough to reproduce. That way, they sustain the chemical process indefinitely. If fermentation engineering with microbes does end up being the disruptive process that supplants animal agriculture, then we should figure out what the expected gain is; hence, onto the next chapter.

CHAPTER TERMS

- **kinetics:** how fast a chemical reaction (conversion) occurs; typically a function of the reactant compound concentrations (e.g. concentration of hydrogen and nitrogen gas for the Haber process)
- **reactors:** environmentally-controllable vats for conducting a chemical reaction. A **bioreactor** controls biochemical reactions especially ones depending on the autocatalysis of microbes.
- **catalyst:** anything that increases the kinetics of a reaction without being consumed itself
- **adsorb:** when a chemical molecule sits on a surface
- **allied products:** parallel products sold by the animal-agriculture industry that are not for direct consumption. These include leather, pharmaceutical products, and the collagen/gelatin in beauty, medical, and food products.
- **denature:** the tendency of proteins to deform under non-native conditions
- **enzymes:** proteins within the biological organisms that function as catalysts for metabolism. Enzymes facilitate the conversion of sugar to ethanol, to CO_2, and to more yeast mass/cells.
- **linear:** a trend or growth that occurs in a straight line over time

- **exponential:** a trend or growth that bends upward over time
- **autocatalytic:** the property of something to make more of itself *faster* (exponentially). All biological organisms are autocatalytic under the right conditions.
- **centrifugation:** A separation process where a liquid sample is placed in a fast rotation chamber to increase gravity and thereby separate the small components from large ones

CHAPTER SUMMARY

Processes ultimately determine what sort of common products come into our lives. For example, the Haber process enabled the facile production of ammonia, which thereby resulted in a population explosion in the last century as it became easier to produce more food with less effort. Likewise, the Haber process, together with refrigeration technology, and general increase in demand for meat, ushered in more efficient animal-product processes, primarily using cows, pigs, and chickens. Looking at the big picture, animal products provide protein—both nondescript and specialized fat, vitamins, and taste profiles. There is no fundamental, physical reason why these molecules cannot be eventually sourced from microorganisms such as yeast and bacteria. These microorganisms exhibit impressive process features (e.g. autocatalysis), and they have, accordingly, endured since ancient times as a means to produce beer, wine, and bread. Furthermore, fermentation-engineering principles enable scalable chemical production using biological systems.

CHAPTER FOUR

Animals by the Numbers

STRUGGLING WITH SCIENCE FICTION

When I was a teenager (the year 2000), the coolest movie in theatres was *The Matrix*. My friends and I discussed it *ad nauseam*, making references to the blue pill/red pill conundrum, and I remember seeing both sequels on the first day each came out. Since it has been twenty years since release, I will feel free to spoil pertinent plot elements to make my point. *The Matrix* follows a computer programmer who finds out that perceived reality is actually a computer simulation, otherwise known as the Matrix. The real world is actually under the control of advanced machines that have conquered the world. My thirteen-year-old self was blown away by the big reveal: humans had been enslaved, sequestered into amniotic pods, and entertained into submission. Humans were necessary to provide energy to the machines, as the atmosphere, post-machine domination, was so polluted that sunlight could not reach the surface. The latent heat from the human bodies powered the machines, while their consciousnesses were under the spell of the Matrix to quell any potential rebellion.

At age thirteen, the twist lingered with me awhile, but for anyone who has a modicum of scientific intuition such an idea becomes absurd after scrutiny. As a much wiser and more well-read adult, I reflected back on *The Matrix*, and my train of bemused thoughts ran like this. "How exactly do humans provide energy? Heat? But you still have to feed the human somehow. That seems ridiculously inefficient. Why not burn the food directly (i.e., not pass it through the humans first)? If these machines are so smart, why aren't they using something more efficient and manageable such as nuclear, or hell, even coal."

Animal technology evokes a similar consternation. When evaluating a cow as a bioreactor that produces meat, milk, and leather, the absurdities are obvious. It takes years for a cow to fully grow into an adult, and the amount of hay we feed the cow dwarfs the amount of food and goods the cow produces. Surely, we can do better, but first we have to understand the physical design and limitations of cows, and animals in general. We will see how these fundamentals limit the overall process metrics because there is no way to engineer animals to overcome the confines erected by our physical world, specifically the inherent limit of chemical diffusion; we must instead turn to new technology with a higher ceiling.

ONTOGENETIC DESIGN

In the last chapter, I extolled the role of processes, such as fermentation, and highlighted how consequential they were to the course of human history. I highlighted that these processes, as any other technology, often displace an inferior, competing process for a given problem. This routinely happens, and I expect it to continue. However, what exactly determines when a process is better? More specifically, what parameters can we calculate in order to compare one process with another? This is necessary in order to compare animal technology to any other that would replace it.

We can intuit some obvious features that make a process better. In the case of a wood-fired engine, we can guess that it is rather inefficient as too much of the wood becomes ash instead of energy. In contrast, we would expect a gasoline engine to be more efficient, as far less solid waste results from its combustion. Stated another way, we get more output per input of gasoline versus wood because gasoline has a higher product **yield**, in this case, the product being energy. So, if wood and gas cost an equal price for the same amount, we would be able to travel further per dollar with gas.

In the Haber process, after the nitrogen and hydrogen gas flow over the catalyst bed, only about fifteen percent of the input nitrogen and hydrogen are converted into ammonia.[93] To circumvent this poor yield, Haber literally circumvented the gases, passing the feed gases through again, recycling them continuously over the catalyst bed. Overall, the process could achieve a yield of ninety-seven percent, meaning an incredibly high total of the input nitrogen and hydrogen would form into the ammonia. In terms of yield, we would be hard pressed to find a superior process to Haber's.

Getting back to cars, we also know that gas engines run hot. Car engines must ignite the gasoline in order to liberate and actualize the energy. Some of the energy expands the air within the closed cylinder forcing the piston down and thereby turning the crank. Much, however, is lost in the form of heat, so there is an inherent limit to the energy yield based on the current gasoline engine design. Maximum theoretical yield from gasoline-powered engines is often calculated in thermodynamics classes by science majors in college. Due to the loss of energy as heat, at best a paltry thirty percent of the total energy in gasoline can be converted to automotive energy to propel the vehicle further. In contrast, electric vehicles do not require heat generation in order to harvest mechanical energy. As a result, far more of the input energy is spent propelling the vehicle forward. Indeed, electric vehicles can use more than fifty percent of the input energy purely for locomotion.[94]

Similar to gasoline engines, incandescent light bulbs generate heat, which undoubtedly saps a chunk of the input energy, which is wasteful, especially when we only desire light. The input energy—an electric current—flows through a filament of tungsten, consequently heating it up. Only when the tungsten reaches a certain temperature does it evaporate and emit the light. In terms of yield, *only five percent of the input energy* is converted into light energy in incandescent bulbs.[95] By contrast, light-emitting diode (LED) bulbs work through the physics of electroluminescence: as electricity courses through LEDs, electrons traverse from a high energy region (the conduction band) to a lower energy region (the valence band). Energy is lost as electrons make this jump, but this loss is our gain, because some of the loss comes in the form of light. LEDs certainly lose energy, but the overall yield is *5 to 10 times that of incandescent light bulbs*. Users of LED bulbs also notice that they run cooler compared to incandescent bulbs.

Indeed, incandescent bulbs carry a fundamental flaw—the release of light is *coupled* to heat generation. Even if we only want light, heat is always accessorized with it. As a result, the best yield possible with incandescent bulbs is, ultimately, capped because we'll always lose some of the input to this heat. This fundamental design flaw of incandescent technology guaranteed its eventual displacement, or limited its ceiling as a technology, to use parlance from Chapter 2. Ultimately, the death of incandescent technology has materialized seemingly in LED technology, which relies on the electroluminescence instead of filament evaporation. Any lightbulb technology with a better inherent design would have ultimately surpassed the yield of an incandescent.

In general, we always desire more output per input, and this is certainly true with food. We crossbreed and cultivate crops so that consumable bits are continually bigger and fatter. Earlier, we discussed how the bananas we eat today are significantly different from ancestral "natural" bananas. Generations of banana breeding has taken place in order to make the yield better; i.e., make more of the banana consumable, make bigger bananas, and make the seeds smaller. For the input to cultivate one tree, growers desire more banana flesh. And certainly, meat producers seek the same for animal products.

However, like the incandescent light bulb and wood-fired engines, animal technology carries a major flaw fundamental to the inherent design. All animals grow **ontogenetically**: they first grow quickly and then at maturity rapidly slow in growth. For anyone who has reared a pet, animal, or child, this is obvious. And for those who are observers, we may intuit that after they're born, they experience an explosion of growth as an infant, a short period of time relative to the rest of their life (**Figure 7**). In the cases of pets, such as a dog or a cat, we know that within about one to two years, our companion will reach full size. They will remain this size for the rest of their lives; hopefully, at least ten healthy, happy years.

In the case of humans, we're generally happy to no longer put on mass after a certain point, but for animal agriculture this is an adverse outcome. An animal agriculturist would much prefer an animal that enlarges continuously because the biomass holds the protein, fat, or molecules for eventual sale. Instead, the animal's growth slows and then stops after a certain point. Even worse, the animal still needs to eat. The consumed food now, however, is no longer fueling the creation of biomass; instead, the nutritional input merely maintains that biomass (**maintenance**).

Figure 7. Ontogenetic growth. A puppy becomes adult-sized in a short time (one to two years) relative to the rest of its life (approximately twelve years).

There is no avoiding this ontogenetic development program. Producers will have to wait until a certain minimum number of their livestock have reached sexual maturity and reproduce; otherwise, they run an unsustainable process and completely forsake the advantage of autocatalysis with the animals. Good business means that only after the animals have reproduced does it make sense to slaughter and harvest their bodily material. Even for many cases where the animals aren't killed for products, the animals must reach reproductive maturity. Heifers must be impregnated in order to provide milk. Hens must be sheltered and fed to be able to produce unfertilized eggs. Therefore, there is no way for animal product producers to avoid the maintenance tax, which becomes most significant as animals reach productive potential.

For any biological system, there is a required input. All life requires water and basic chemical elements to thrive, including carbon, nitrogen, phosphorus, etc. This input is chemically converted by the biological system, using the enzymes within, into the two rough categories of products: biomass and maintenance.

Maintenance refers to the process of renewal and replacement of dying cells, as well as the generation of energy so that animals can walk, run, and breathe. Metabolic maintenance is analogous to a car running on gasoline or a LED display depleting a battery. Biological life has evolved in such a way as to execute mental procedures, find companions, experience emotions, and perform physical actions powered by the food consumed.

All cells within an organism will have a maintenance requirement, meaning that a minimal amount of nutrient (e.g. energy) per time is needed to maintain cellular function. If the nutrient input exceeds the cellular maintenance amount, then the excess nutrient can be used to make more mass. This is why younger animals and humans can grow so fast; they have fewer cells and therefore can divert more of the food input toward making new mass. Eventually growth slows as more biomass accumulates and the maintenance payments only enlarge. We settle in at a set size when the maintenance requirements equal the input.

These maintenance requirements are more costly for ontogenetically-growing organisms compared to plants and smaller microbes that follow a different growth trajectory. Therefore, we can cast ontogenetic growth as the incandescent light bulb, while a bioreactor growing microbial meat is the superior LED design. Animals inherently grow this way because of the inherent physics of how they distribute nutrients within their body. This distribution is non-optimal for production purposes with a subsequent, significant drop in eventual yield.

YIELD

Ontogenetic design dictates organisms of a certain size, a size selected per the evolutionary niche and objectives. For most biological matter, transporting and delivering nutrients to all cells remains the fundamental challenge imposed by the physics of the universe. Our cells demand nutrients at a requisite rate; therefore, we need networks and organs to actively purvey sustenance. Consider oxygen, which is required by nearly every cell for viability. After entry into the lungs, oxygen binds to the hemoglobin protein in our blood, and the oxygen subsequently circulates around our bodies, vitalizing our organs and tissues. Likewise, we transport hormones, sugar, and waste through blood. When our fight-or-flight responses kick in, adrenaline pumps into our blood, quickening the heart rate and making our palms sweat—all in under a minute, thanks to our efficient blood circulation system.

Smaller organisms, such as insects and mollusks, do not possess such sophisticated transport systems. They manage with simply an open circulation system: one artery surrounded by a giant soup of goo. Insects do have hearts to slosh the fluid around, but their circulation systems lack the fractal patterning characteristic of humans and animals, where arteries run into arterioles then into capillaries.[96] Nonetheless, nutrients reach the extremities, and fight-or-flight hormones function. An open circulation system satisfies the basic requirements of the organism, though the system caps its maximum size; as evidenced by large mollusks such as squids and octopi that also run on closed circulation systems. Stated another way: size dictates the type of circulation system an organism requires.

Resource-wise, pumping fluid throughout our body (or the body of an animal) is costly. A dedicated organ, the heart, beats rapidly, pushing blood through the circulation system of veins and arteries continuously. The energy required to move liquid throughout the body is a prime example of maintenance energy. Furthermore, we know that our circulation system works at roughly a constant flow rate of about five liters per minute. Provided that we remain healthy, we maintain this rate. More importantly, our flow rate does not increase as we age. And given that this flow supplies the nutrients in our body, we cannot grow in an even pattern, or linearly. For us to be able to grow linearly, our circulation system would actually *have to speed up over time* in order to maintain our growing mass while adding more to it.

The concept of circulation is further complicated because, according to Bernoulli's principle, which is a good mathematical model of our blood flow, we do not receive a flow rate increase directly proportional to the amount of work put in. If our heart pumps twice as hard, i.e., creates double the pressure gradient, we do not achieve two times faster flow. We only receive about 1.4 times more. In terms of energetics, it'd be evolutionarily nonsensical for us to continually enlarge after a point. Therefore, we can also think of the blood-flow rate and the circulation system as capping our size.

As we inferred, the maintenance costs only go up during development, as we have more mass taxing the nutritional income stream. As a result, animal and human development curves downwards and flattens at the average size of adulthood (**Figure 8**), conforming with the experience we witness in puppies, kittens, and children. Therefore, organisms that do not rely on circulation systems may be able to grow faster (and bigger) because they're not shackled by the constraints of a fluid-flow system. Imagine instead that the organism was miniaturized, and the nutrients could be supplied entirely by **diffusion**. Diffusion is the latent tendency of molecules to disperse and move around within a space or a medium. We can imagine a scented candle filling the room with aroma. We don't have to fan the candle as the scent molecules diffuse throughout the space of the room freely. So even if the candle is on the other side of the room, we still smell it. In the same way, tea molecules seep out of the tea bag, creating a visual black cloud suspended in our cup. If we wait long enough, that cloud will envelop the entire liquid without stirring or moving the bag.

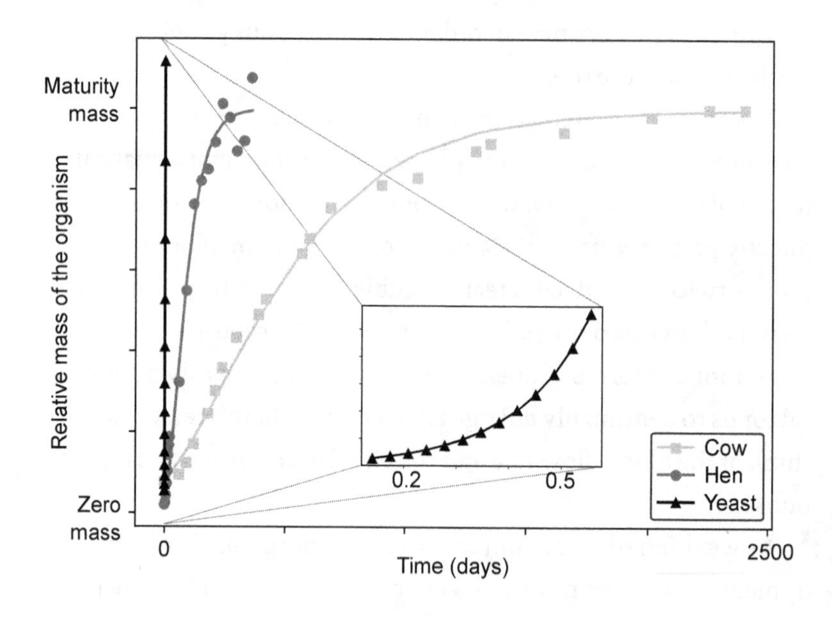

Figure 8. Time to maturity is longer for bigger organisms. The time to maturity in different organisms, from their birth to adulthood. The inset is for yeast, and magnifies the relevant part of its otherwise compressed growth curve.[97]

We know that if the room is big enough or if we're outside, then the candle scent may be less detectible as it has a specific range. Likewise, we intuit that the tea bag's cloud will only get so far in a larger container of water. Therefore, as we shrink down the size—smaller than what we can see by eye, we get to the world of microorganisms. For microbes (microorganisms), diffusion reigns supreme as the way to supply nutrients. These creatures do not need a circulatory system. They are so small that oxygen can reach one end to the other in two to three microseconds; or stated another way, can bounce back and forth from one end to another roughly half-a-million times in one second, all by diffusion.[98] Accordingly, we see that yeast, which receives its nutrients via diffusion, shows an upward development trajectory (exponential), as indicated by the inset graph. Without baggage such as a circulation system, these organisms are able to devote more of their input resources to building mass; they do not incur the maintenance costs associated with an energy-intensive circulatory system.

The influential team of Geoffrey West, Brian J. Enquist, and James H. Brown (WEB) have greatly contributed to the mathematics behind modeling how metrics of organisms change based on their size,[99] known as the field of **allometry**. Should you find the topic of interest, I highly recommend West's excellent book *Scale*.[100] From WEB's equations, I calculate the expected yield for all animals is fourteen percent.[101] This means that only fourteen percent of what the animal consumes is converted into mass by the time it matures (and is presumably harvested and slaughtered) while everything else supports maintenance.

This fourteen percent value also presumes that all of the mass of the animal is used. Clearly, we're overestimating there, as we generally don't use animal brains, feathers, or intestines. They're all waste unless the animal agriculturist can wring out more allied products from them. Even so, this would add but a trifle to the eventual yield. Furthermore, this same percentage applies to any of the species that adhere to the model posed by WEB, including hens, pigs, cows, salmon, cod, shrimp, and rabbits.

On the other hand, organisms that do not adhere to the animal ontogenetic model, such as plants and microbes, have vastly improved yields. For plants, the yield (how much input carbon dioxide they turn into mass) hovers generally around fifty percent, and that includes lettuce,[102] grass,[103] alfalfa,[104] and trees.[105] For microbes, yields are even higher—generally above eighty percent—and some species reach over 99.5 percent.[106] Stated explicitly, it's as if animals as a food technology are akin to the fundamentally limited gas engines or incandescent bulbs, and microbes are like the electric engines or LEDs.

To showcase how amenable microbes are to potentially high-yield processes for desired compounds, let's look again to yeast. The yeast species *Komagataella phaffii* grows well in a bioreactor and can be designed to secrete heterologous protein. So, if we introduce the gene for casein protein and get the modifications correct, then our process would obliterate any cow-based one in efficiency. The yeast could be cultivated in a membrane reactor (**Figure 9**), in which yeast cells are contained by membranes that only permit the passage of liquid and free protein. First, we would grow our yeast, then we would start the flow of a fresh medium from which the yeast would make casein. The newly created protein would be carried in our flow and could be collected downstream of the membrane. This process could be carried out indefinitely. Certainly, some yeast would die but could be replenished readily. We'd easily surpass the yield limit of a cow-based process when we've developed the *K. phaffii* technology to sufficient fruition.

Clearly, we should be avoiding large animals given the physical constraints of nutrient transport. We're better served to stick with smaller ones that efficiently convert what they eat into mass. Additionally, yield is not the only metric in which animals fall short. The *speed* at which animals can grow is also fundamentally limited, already apparent in the development curve presented in **Figure 8**.

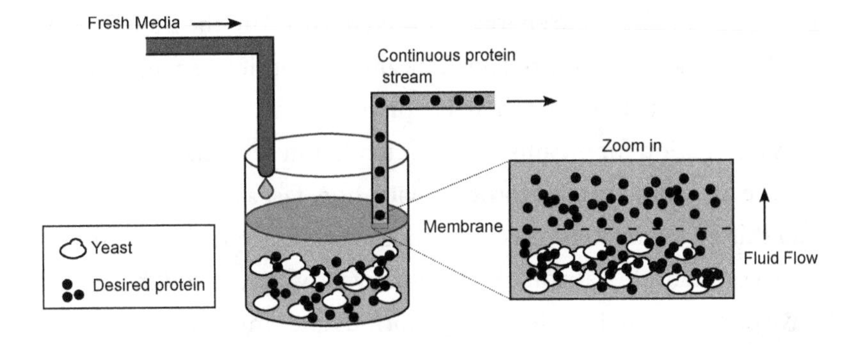

Figure 9. A schematic for a potential high-yield protein production process using the yeast, *K. phaffii*. Fresh media feeds the reaction where the yeast cells produce and secrete protein. The protein separates from the reaction by free diffusion through a membrane and can be pumped out.

PRODUCTIVITY

Earlier, we discussed the five-liters-per-minute blood flow rate of the human body and how costly it would be to accelerate the speed of the fluid circulation. Cells cannot grow faster than the nutrient supply dictates. For example, if I consume two thousand calories a day, those calories can only build so much mass. Forming 4.5 kg (10 lb.) of body mass from those calories would be impossible per the law of conservation of mass, but forming 0.02 kg (0.05 lb.) may be entirely feasible. If we miniaturize this idea and consider how many calories each cell is receiving from blood flow, then we should conclude there is a limit to the exact amount of mass that each cell can put on per day. Stated another way, given the necessity for a circulation system to carry nutrients, the *growth speed* of organisms is **limited** by how quickly the flow can supply the nutrients. Furthermore, the diminishing returns of Bernoulli's principle means that proportionally more energy is required to increase the speed. Therefore, we can imagine an ultimate limit to the speed of fluid flow where the energy required to speed up the fluid flow exceeds what's possible to consume and convert into energy. We ineluctably conclude that there is a physical limit to the *speed* of mass formation for organisms with circulation systems.

We care about process **productivity**, how fast a process occurs, and this can ultimately affect the price, even if the yield is impressive. For example, saffron, the fragrant, yellow spice thought to have originated in Iran, is highly laborious and time-consuming to produce: anywhere from 50 thousand to 150 thousand flowers are required to render 1 kg (2.2 lb.) of saffron.[107] The spice comes from three thread-like protrusions from the center of the flower. These threads must be plucked out manually and dried. This process must be achieved by hand as the threads are delicate and liable to disintegrate with automation. Saffron production remains slow and laborious, and that ultimately affects the price. Ten thousand dollars per kilogram is not uncommon for saffron, about a sixth of the price of gold.

If the process to produce saffron was faster, the price would surely be lower as the labor costs alone would drop, and the amount of space required would diminish. Thankfully, most cooking does not require 1 kg (2.2 lb.) of saffron. Most of the time, just a pinch does the trick.

A more cautionary example of the power of productivity is found in the potato's fate in Ireland. Ireland imported potatoes around the late 16th century.[108] Potatoes became popular throughout the country for both yield and productivity reasons, as with agriculture, both are often intertwined. Growing potatoes requires a requisite amount of land, generally the mathematical input of the yield calculation. With potatoes, farmers could reap more food per unit of land. Another way to look at it is that the potato grants more food per hour spent in the field for the same amount of land. This is productivity when seen specifically through the time dimension.

Given that potatoes are superlative in both yield and productivity dimensions and amenable to growing in a cooler climate, they quickly dominated the agricultural landscape for Irish farmers in the 19th century. Also, compared to competing crops, they were more nutritious. By 1840, Irish farmers were eating three potatoes per day.[109] This overdependence on potatoes soon led to deleterious consequences a few years later when a disease called the potato blight arrived in Europe. It devastated the Irish potato production. In 1846, production of potatoes dropped approximately seventy-five percent from the two years prior due to blight infestation, which rendered potatoes unfit for consumption.[110] Naturally, this destroyed the Irish population. An estimated 1.5 million people died from hunger or related causes, including dehydration from eating blighted potatoes and suffering from lethal diarrhea.

The process metrics of the potato helped it monopolize the Irish agriculture scene, unintentionally also priming Ireland for devastation. Animal agriculture producers have also deeply cared how quickly their animals grow and made concerted efforts to speed up the growth of production animals. As Smil highlights in *Should We Eat Meat?,* the wild progenitor of the modern chicken is the red jungle fowl indigenous to South Asia. This bird reached sexual maturity twenty-five weeks after birth and required six months to reach full weight.[111] In contrast, modern free-range chickens reach slaughter weight in fourteen weeks, and chickens in confinement reach slaughter weight in a mere six weeks while weighing fifty to sixty percent more than the free-range counterparts.

These metrics have similarly held for other livestock animals. In confined operations, female pigs are ready to reproduce a mere thirty-two weeks after birth, in contrast to two years in the wild.

With developments in raising cows, a 145-week rearing period consisting of fresh range grazing was shortened to 56 weeks. In the meat-industry terminology, the take-off weight metric captures the same information; it quantifies how quickly an animal grows in a set amount of time. A take-off rate of one means that, on average, a given animal in the country takes one year of development time to reach slaughter weight. A take-off rate of two means six months, a rate of four means three months, and so forth.

Producers use selective breeding to continually speed up the growth of the species. As Smil highlights, this progression led to chicken meat supplanting pork as the dominant flesh consumed. Now, the industry breeds chickens[112] and turkeys[113] with such large breasts that they can't walk or have sex. Cows are fatter and eat more.[114] If you seek even more examples, I suggest Jonathan Safran Foer's book *Eating Animals*.[115] It documents many ignominious ways in which breeders have attempted to boost productivity by, for example, getting hens to lay more eggs, and cramming animals in tight spaces. Despite the lengths taken, animals are *still* fundamentally lacking in productivity compared to prospective competitors (e.g., plants, yeast) owing to their self-limiting circulatory systems.

Biomass productivity has another name in the biochemical engineering space: **growth rate**. Growth rates are similar to the take-off metric used for animals, but they're not directly comparable. In particular, microorganisms can have constant growth rates such that, in certain bioreactor processes, their growth rates can be maintained indefinitely. This is absolutely not true with animals as we see in the development curve (**Figure 8**), with tapering growth as the animal reaches maturity weight. To get around this, we can consider the similar strategy that I used to derive the fourteen-percent yield number for animals and consider the weighted average growth rate for animals in order to directly compare to microbes and plants. I've derived such an equation, again from the framework proposed by the WEB team:[116]

$$\mu_{\text{average}} = (0.0077)\,\frac{a}{M^{\frac{1}{4}}}\ \text{[per Hour]}$$

Unlike the yield value, which was constant for different animal species, productivity (growth rate) scales with animals based on two parameters *a* and *M*. The *a* value is the ratio of basal metabolism versus the energy required to make a new cell. Understandably, when this value increases, the animal is able to grow faster. If it's cheap to make new cells, then the animal grows faster. Whereas, *M* is the maturity (slaughter) weight, i.e., for cows it's around 450 kg (992 lb.), per the last chapter. Interestingly, this equation suggests that the bigger an animal can eventually become, the slower it must grow. This squares with what we know about allometric scaling laws; as species get bigger, their metabolic requirements increase less than expected on a per mass basis. For example, a guppy has roughly double the maturity mass of a shrimp. However, a guppy's metabolism runs only seventy-five percent more, not the expected hundred percent. This relationship holds for numerous animals, from ones as small as mice up to elephants.[117] Not all animals adhere to this trend, with humans a notable exception, but it does describe a large swath of the animal kingdom.

Nonetheless, the relationship suggests that evolutionary optimization slows down the metabolism of larger species. This makes sense based on earlier discussion about the balance of nutrients from circulation flow and the speed of the growth. Because it's so costly to hasten fluid flow, evolution selects for slower growth in larger animals. Consider the jittery hare versus the bigger, plodding tortoise. As a result of the metabolic slowdown, large animals are able to maintain the same fourteen-percent yield as smaller ones. But the tradeoff is the longer development trajectories: large animals grow more slowly with everything else being equal (particularly the *a* value) compared to smaller ones; hence, the division by $M^{1/4}$ in the growth rate equation.

I've shown the growth rates across the different kingdoms (**Figure 10**) on a log scale where the values are seen as multiples of one another. For example, the growth rate for bacteria (one per hour) is roughly one hundred times that of the leafy greens category (0.0083 per hour). Earlier we described the wondrous doubling capability of yeast. With a growth rate of 0.3 per hour, a single yeast organism could double itself to the mass of the Earth in eight days, using our nonexistent, celestial-sized bioreactor. If we could do the same with cows, it'd take over 150 years because of how slowly cows add mass to themselves. On average, animals grow a thousand to 10 thousand times more slowly than any microbe.

Consider the possibilities from these striking numbers. Instead of trying to gift goats to third-world countries, as some charities pursue rather blindly,[118] why not give a bioreactor? The bioreactor could start generating food in mere hours and would use a fraction of the resources, given the high yield of microbes. In terms of allaying destitution, we'd be hard pressed to do better, especially on the lower rungs of the poverty ladder, meeting and exceeding immediate and long-term nutritional and water needs.[119] But of course, we lack such a cornucopia to grant because of insufficient scientific and technological development. We'll broach this again later.

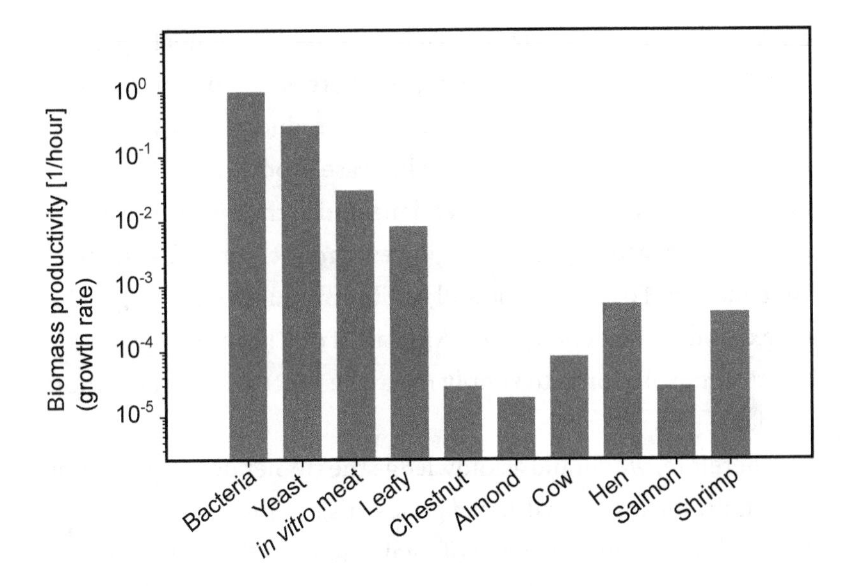

Figure 10. Growth rates across different organisms and biological systems. A representative growth rate is plotted for each categorization of organisms. The vertical scale is log10 based; for example, bacterial growth rate is roughly one hundred times higher than the leafy category, which includes mustard leaves.

I acknowledge that tree nuts, such as chestnuts and almonds, come off poorly in this analysis. I was not surprised by that, but I was surprised by *how* poorly. Trees require a circulation system akin to that of animals to distribute their nutrients, thereby lowering growth rates. The same principles apply, in that smaller plants can grow faster because diffusion imposes less of an energy burden. Plants are also unique in that they turn carbon dioxide within the air into mass. Catalytically, this is an inefficient process, and there may even be fundamental physical limits to this as well.[120]

There's a consequence to the slow growth of land animals in that the terrible animal-based productivity means otherwise arable land is dedicated instead to animal technologies. Animal agriculture has monopolized much of our terrestrial, ice-free surface. Specifically, a whopping thirty percent of such land is used for animal agriculture.[121] And the explanation is simple. There is a lot of demand for animal products. To meet such demand, producers have historically employed contemptible means to increase productivity. For example, battery cages confine and maximize the number of broiler chickens in a facility, increasing process productivity. That alone isn't enough. To counter the still terrible productivity rates, producers expanded their enterprises. After all, if one has a slow process, the current solution is to simply make the process bigger, not more efficient.

Therefore, we should acknowledge the tremendous opportunity cost to producing animal-based goods. It's not just the copious CO_2—about eighteen percent of total emissions[122]—that the animal-industry generates. We also must note all the lost and putative forests that could be drawing back CO_2. Without animal agriculture, we'd almost certainly have more trees and forests, evidenced by clearance activity of the Amazon rainforest for animal agriculture.[123] Switching to a more productive biological substrate (e.g., yeast), would free the same multiple of equivalent land. For instance, if yeast replaced cows for products, we'd free up over 99.9 percent of arable land from the tentacles of the bovine-agriculture behemoth.

LIMITS OF ANIMAL TECHNOLOGY

I can anticipate some dissension, such as: couldn't we engineer or breed animals to have better metrics? We could conceivably engineer a faster-growing, large animal. But a large organism requires a circulation system of some type in order to deliver nutrients throughout its biomass because diffusion alone is unable to get the job done. Also, there is always a tug of war between growth and yield. If we increase the speed of growth (productivity), then we'll sacrifice yield. And the balance of circulation flow and metabolic requirement places an absolute physical limit on how high we can push these numbers.

Inexorably, we must conclude that we *don't* want to breed a large animal because its efficiency of production is never going to exceed that of a smaller organism. We're better off trying to take individual cells and grow them in bioreactors. The surrounding, nourishing liquid media will bathe the cells with nutrients, obviating the need for a sophisticated circulation system and countervailing the limits of diffusion. Such a scenario directly corresponds to the features of microbes undergoing exponential growth and experiencing that growth curve with a trajectory curving upward. In fact, this argument favors ***in vitro* meat**, which involves taking animal cells, placing them in a liquid bioreactor, and reprogramming them so that they can grow directly into a steak. *In vitro* meat is estimated to grow about one hundred times faster than traditional animals, as shown in **Figure 10**. It skips the expensive circulatory system and relies more on diffusion.

Analyses from this chapter should not surprise us because animals did not evolve to be reactors for our chemical processes. Rather, they were evolutionarily optimized to reproduce and survive in their biological niche. The niche for animals is large and vast. Compared to simpler organisms such as bacteria, yeast, and fungi, animals can consume many different types of food and inhabit different environments. Just think about humans; we've prehistorically established and explored areas as diverse as the Australian outback and the Bering Strait—the hypothetical crossing that once connected Russia and Alaska. We are also mobile compared to sedentary species, such as plants and coral on reefs. If we were to come up with a spectrum of how generalist versus how specialized a species is, we'd invariably place animals toward the more generalist end. As a result, animals need to have a variety of functions—from being able to digest and consume varied kinds of food to having stem cells that become anything: hair, skin, nerves, blood vessels, muscles, etc.

This generality is programmed into DNA, which is in every developing cell in animal bodies. Consider human conception: upon fertilization, there is a singular cell, the zygote, that divides itself exponentially, eventually constructing a full human. This zygote contains all the DNA potential to create all the different parts of a human body. This encompassing DNA is passed to each dividing cell. As a result, progenitor cells—also termed stem cells—can develop into many different cell types. It's as if all the cells are carrying backpacks along their development hike. They need to have all the tools in their backpack in order to evolve into everything from a skin cell to a brain nerve cell. Certainly, once a path is chosen, these cells can lighten their load and dispense of some unnecessary baggage, but this generally occurs at the end of development once the animal or human is fully grown. Furthermore, the cells can't dispense of everything in the backpack; for example, they must respond to many kinds of hormone signals and so need to retain that processing equipment.

In contrast, the aforementioned microbes are much more specialized and designed to grow only in environments with a narrow range of temperature, pH, salinity, in particular kinds of nutrients, and at a given moisture level. Outside of these environments, these organisms will remain either dormant or simply die. But within their respective environments, these critters flourish at the eye-popping growth rates that I've highlighted earlier. They are optimized to their narrow niche—excelling in defined environments.

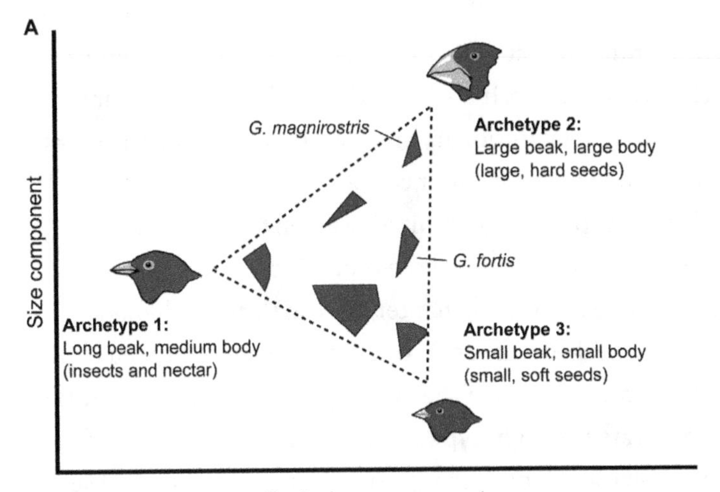

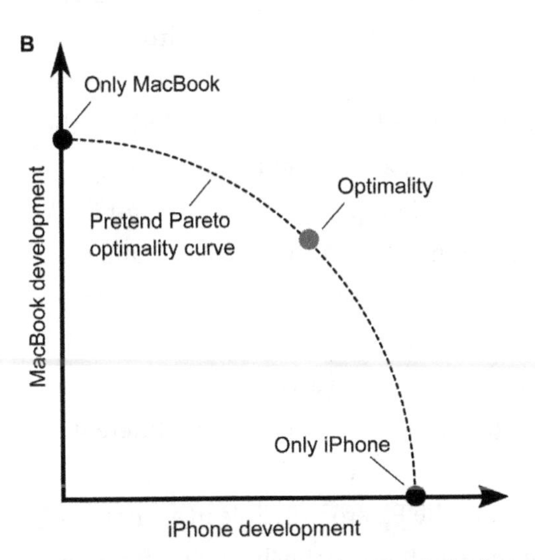

Figure 11 A and B. The Pareto frontier in biology and business. (**A**) All of Darwin's finches fall on a triangle with vertices defined by three archetypes. The interior polygons indicate finch species native to different Galapagos islands. Two examples are shown: G. magnirostris and G. fortis. From Shoval et al. 2012.[124] (**B**) An instructive, but fake Pareto curve for research and development expenditure at Apple, Inc. The dotted line (Pareto curve) indicates the amount of development Apple can perform for both the MacBook and iPhone simultaneously. Presuming a constrained amount of resources, optimality occurs at the center of the curve, if the total development is to be maximized (development of the iPhone plus development of the MacBook).

The more functions at which a species must excel, the worse they perform at each one. It's a "jack of all trades, master of none" situation. In fact, in evolutionary analysis, this is termed the **Pareto frontier** (**Figure 11**). This frontier can be visualized as a confined area for different species to exhibit for a given trait, and the area represents the physical limits of the traits. For example, owing to the finch example in the first chapter, we can think of this as an axis with large-beaked finches on one side and small-beaked on the other. The terminal points of the axis, representing the extremities of beak length are the **archetypes** or exemplar representatives of a specific beak size and length group. Suppose that the types of seeds constantly varied between hard ones and soft ones, and, further, that this was a daily challenge. In such a scenario, the fittest species might not be the large-beaked archetype or the short-beaked one; instead, it might be an intermediate between the two. Indeed, the lab of systems-biologist Uri Alon has highlighted this principle in Darwin's finches, with a third archetype in the mix, a long-billed finch that feasts on insects and nectar.[125] Alon's lab has found that eight different species, each one endogenous to a specific Galapagos island, fit beautifully into this triangle, the Pareto frontier for the finch bills. Specifically, the triangle could account for ninety-nine percent of the beak and body size metrics for all of the different finch species (**Figure 11A**).

Pareto frontier comes from the Pareto principle in economics. This principle states that there will invariably be tradeoffs when allocating resources to one effort versus the other. We can think of a company, for example Apple, which has both its iPhones and its Mac computers. Apple has a finite budget for research and development, and it can certainly place all of its resources toward development of one or the other, but it will likely go for an intermediate solution. Apple does fund efforts that further *both* iPhones and Mac computers, for example with shared software. So, a Pareto curve may not always be a straight line (**Figure 11B**) and will actually curve out when synergistic efforts are present, as suggested in the figure. And presumably, Apple goes for the point on this curve that maximizes its profit to the best of its forecasting abilities.

Alon's lab did not stop their analysis with finches but also examined bacteria[126] along with the labs of Terence Hwa,[127] and my postdoctoral advisor, Uwe Sauer.[128] The researchers noticed that bacteria would divert resources inside their cellular bodies toward a given situation. When a lot of nutrients were available, the bacteria would allocate more internal resources toward growing as quickly as possible. When awash with nutrients, the fundamental limit for how fast bacteria can grow is based on their ability to synthesize protein. Therefore, when a windfall of nutrients presents itself, the bacteria would create more protein-synthesizing **ribosomes** so that they could crank out protein as fast as possible. However, when resources are scant, bacteria are best served committing more resources to scavenging fleeting nutrients; otherwise, they risk being outcompeted by more voracious competitors. Instead of ribosomes, the bacteria will synthesize internal proteins that grant speedier nutrient-uptake for such conditions. These bacteria, *Escherichia coli*, inhabit our gut. None of us eat continuously; instead, we eat meals and snacks, and appropriately our gut experiences these cycles of periods with high levels of nutrients interspersed with periods of starvation. Accordingly, the *Escherichia coli* bacteria exhibit a point optimality on the Pareto frontier between the two archetypes, as highlighted by the work of Benjamin Towbin, Uri Alon, and their team.[129]

So, these studies imply that we have not yet found the most efficient production organism because current ones (e.g. *E. coli* and yeast) all carry baggage from evolution in their natural niche.

Ideally, we just want something that produces the desired protein, fat, and molecules as efficiently as possible. Certainly, we'll want to take advantage of the autocatalytic abilities latent in biology. But otherwise, we just need a biological system that satisfies our production objectives. We can dispense with everything else: the circulation system, the bones, the mobility, and most importantly, the capacity for suffering that makes animal production particularly deplorable. Instead, we could conceivably engineer a simple system that solidifies our nutrition and nutrient security requirements. We could gift bioreactors with such production organisms to starving villages in Niger and Afghanistan, and farmers could reap sufficient food in a trivial amount of time.

MODERN FERMENTATION-DERIVED MEATS

The numbers and fundamentals I've presented also suggest that there's a considerable financial opportunity for entrepreneurs and industrialists to exploit in the shift from animal products to fermentation-derived goods. So where is the gold rush?

In mid-2020, a state of the fermentation industry report was published by the Good Food Institute (GFI), the non-profit advocacy group and funder of science-based innovation with a mission to replace animal products.[130] The report highlights sixty-eight companies using or pursuing microbial fermentation to replace common animal products. Some are focused on specific proteins, such as Clara Foods, which is producing egg albumin using yeast, i.e., the specialized case mentioned earlier. And the report also highlights twenty biomass-focused companies, producing meat that is purely bioreactor-derived material like Quorn.

All of these biomass companies, with the exception of Monde Nissin, the company that produces Quorn, were founded in 2013 or later. Four-fifths were founded in 2016 or later. There is a gold rush;

it's happening right under our eyes. And we can expect more companies to enter the fray. The GFI report highlights the expanding investment—more money was invested in microbial fermentation (about $435 million) to replace animal products in just the first half of 2020 than the previous year, which itself had more fermentation investment than any year before.

The microbial fermentation companies are well attuned to the production advantages of their technology. Meati, one of twenty biomass companies, like Quorn, produces faux chicken and steak from fungal mycelia. Meati echoes notions presented throughout this chapter directly on their website: "Our mycelium becomes 'full-grown' overnight...if you can believe it. In case you didn't know, a cow can't do that."[131]

The Quorn fermentation process can double its mass every three to four hours;[132] that's almost as good as yeast in **Figure 10**. Yield is a work in progress at thirty percent, mostly due to the additional step of removing the excess nucleic acids that cost the process a whopping thirty percent in efficiency.[133] There is room to improve this separation though.

These companies and this technology are limited by consumer acceptance. As we saw with Quorn, a pharmaceutical-level testing and approval process was needed before it was deemed safe to eat. Furthermore, microbial fermentation would be greatly enhanced with access to genetically modified organism (GMO) technology. In the next chapter, we'll further examine genetically modified technology as a potentially decisive tool in this quest, particularly getting over the misplaced wariness of it.

CHAPTER TERMS

- **yield**: how much of a desired output versus input a chemical process achieves. For example, two cups of flour may yield two dozen cookies.
- **biological maintenance**: energetically-demanding activities within a biological organism that are essential for survival and growth but do *not* result in additional biomass
- **ontogeny**: the development of an organism from infancy into adulthood
- **animal ontogenetic model**: a flattening curve describing the development of animals. Infants grow quickly but slow down later, due to maintenance costs, and eventually stop growing, as do all animals.
- **limited**: when discussing a process, the speed (productivity) will always be bracketed by specific physics. We, therefore, can say that a certain process is limited by something.
- **productivity**: the speed of a process. For example, my kitchen and I have a productivity of twenty cookies per hour.
- **growth rate**: how fast an organism adds biomass to itself; biomass productivity
- **diffusion**: the tendency of molecules to disperse, particularly in fluids. To some extent, diffusion supplies nutrients for all biological life; however, diffusion alone is not fast enough to sustain life beyond the microorganism domain.
- **allometry**: the study of biological metrics (e.g. metabolism, heart rate) as a function of organism size
- ***in vitro* meat**: taking a stem cell—say from an animal—and growing that cell in a bioreactor into tissue

- **Pareto frontier**: the "ceiling" or limit of a biological species. Once organisms reach their Pareto frontier, minimal evolution can be expected unless their niche (environment) or objectives (e.g., domestication versus wild) change.
- **archetypes**: variants at the vertices of the Pareto frontier for a biological trait/phenotype. They serve as an idealized representation of what's physically possible for a certain trait, such as growth rate or beak length.
- **ribosomes**: cellular machines comprised of RNA and protein used to manufacture protein

CHAPTER SUMMARY

Animal technology is ultimately limited by its inherent design. In order to grow so large, animals require an intricate circulatory system so that nutrients may be delivered to all parts of their bodies, and waste can be shuttled to exits. Thus, this design limits both the yield and productivity of animal-based products. The food input that animals consume must partially build and maintain these auxiliary features, subtracting the input going directly toward the animal mass. Furthermore, this design caps the speed at which animals can grow. In contrast, microorganisms skirt the limits of diffusion when small enough. As a result, they have vastly better yields compared to animals and other larger organisms. Furthermore, they are capable of growing orders of magnitude faster under the right conditions—ones that we can control via bioreactor processes. Knowing all these features, we must conclude that a microorganism-based process has a fundamentally superior design compared to an animal or plant-based one. And appreciating how processes are technologies that are liable for replacement, we conclude that with enough development, micro-organism technology will generate most, if not all, of our food.

CHAPTER FIVE

Intractability of Animal Technology

SYNTHETIC BIOLOGY

When I started my doctorate, the most exciting field was synthetic biology. In my program at Northwestern University in 2010, there were approximately three synthetic biology positions available, and seemingly around twenty students vying for the spots. I felt extraordinarily fortunate to earn a ticket to board the synthetic biology bandwagon. After all, it was thought to be the future of science, along with fields such as nanotechnology and quantum computing. Synthetic biology was to be the means to generate a fully renewable economy and enable us, for example, to successfully settle on Mars.

Synthetic biology entails creating biological systems that don't exist in nature. This could look like engineering bacteria that produce biofuels from renewable sugar stocks. Or engineering dry baker's yeast to detect diseases in patient samples in areas without refrigeration access. It could even be engineering a person's immune cells to attack cancer in the body. The field initially burgeoned as a result of cheap DNA synthesis and sequencing (what an economist would characterize as technology lowering the barriers of entry to the market). Both ideas are central to synthesizing novel biological entities. Advances in the field continue to gain momentum thanks to more facile genetic manipulation techniques, such as the famous CRISPR-Cas9 system, mathematical modeling, instrumentation, and the biological understanding that underpins the engineering efforts.

My doctoral advisor proposed that I use protein degradation for improving biological-based chemical production using synthetic biology methods. Specifically, we could target a protein that enabled cell growth. Once the cells were grown, the protein could be degraded away leaving an intact, viable, and chemical-producing cell—say synthesizing a bioplastic. We can imagine the overall formula looking like *Input →Cell growth+Bioplastic*. So naïvely, if we degrade the growth protein, our equation should reduce to *Input →Bioplastic*. By subtraction of cell growth from the output, the yield for the bioplastic should dramatically increase.

Despite the ostensible glamor and theoretical ease of synthetic biology, the actual work was a lot of trial-and-error. First, I designed a DNA sequence responsive to the chemical inducer followed by the gene coding for an unmasking protein. According to this design, the unmasking protein should only be synthesized when the bacteria were presented with the chemical inducer, as per my intention. Once synthesized, the unmasking protein would then interact against the target growth-enabling protein, revealing a signal for that enzyme to be degraded by the cell. An analogy here might be that the unmasking protein is a bounty hunter, finding and apprehending the bail jumper—our target protein. The bounty hunter ultimately hands over the fugitive to the judicial system—the degradation machinery.

After the DNA design, I purchased some of the sequences partially synthesized, but then I would have to assemble the final DNA construct myself. Making the DNA itself was a notoriously fickle process in that, sometimes, I would generate a nonfunctional piece of DNA and unintentionally introduce errors as I synthesized or stitched DNA together. Often, I didn't get the behavior I wanted; for example, the sequence of DNA that responded to the chemical inducer might actually be leaky, rendering an unreliable unmasking protein. I would often spend long days and weekends hoping that the latest iteration would be the ultimate point. Cruelly, there would often be even more unintended, unforeseen consequences. For example, degrading target proteins clogged the cells' protein degradation processes essential for a healthy, fast-growing cell. Even when the system worked well, the cells plodded along, likely unsuited for a prime-time chemical production. Overall though, this modus of conjecturing a DNA sequence, constructing/introducing it, and testing it is not new or exclusive to synthetic biology; it is the **design-build-test** cycle paradigm.

As with nearly any engineering discipline, synthetic biology demands design-build-test cycles. We designed the DNA constructs with a conjectured behavior, then "built" the DNA using a third-party synthesis company that supplied the material (or synthesized it ourselves in lab), and then finally we tested the performance within the bacteria. In this approach, when something does go wrong, we likely must make another conjectured DNA construct and test that. However, as mentioned, this modus has been applied in other engineering efforts: the principles of the light bulb were well-founded before a tenable mass-producible product could be developed. Thomas Edison's team tried thousands of light bulb builds before achieving a suitable one.[134]

It can be an incredibly discouraging, prolonged process that feels more like an art than science. Synthetic biology is still far from the modularity of, say, building circuits with all the necessary components, long ago perfected and available "off the shelf." The resistors and operational amplifiers of synthetic biology can affect one another due to the Pareto frontier and evolution, as we'll discuss in this chapter. For these reasons, I'm not confident that synthetic biology will reach the engineerability of circuit design. Design-test-build might be the only way.

As a synthetic biologist, I am modifying DNA, and this practice does stoke fear in the public. I remember when I watched *Jurassic World*, the soft reboot of the seminal film *Jurassic Park*. The movie's main antagonist, *Indominus rex*, an engineered, super dinosaur, clearly plays to the fears about **genetically modified organisms** (GMO). Using GMO technology, *I. rex* was a biological portmanteau of different species—*Tyrannosaurus rex*, cuttlefish, tree frog, viper snake, etc.—created to achieve a terrifying, highly competent villain. Synthetic biology entails genetic engineering and creating genetically modified organisms, but I sincerely doubt that you'll find the term GMO used on any synthetic biology website. However, the absence of the term can't deny the plain fact: the processes of synthetic biology and genetically modifying organisms are the same thing.

Despite the difficult methodology and scattershot outcome, I see GMO technology as largely beneficial in the effort to supplant animal products, and it's in everyone's ultimate interest, including animal-rights activists, environmentalists, synthetic biologists, and anyone else to emphasize the positive value, problem-solving capability of this technology. I believe the fear is rooted in the naturalism fallacy I discussed in Chapter 1. My hope is that with better understanding of the topic, especially the limits of genetically modified organisms, humanity will better appreciate and promote this technology.

THE COMPLEXITY OF BIOLOGY

Jurassic World's *Indominus rex* was the result of DNA splicing from a variety of sources: velociraptor DNA for increased running speed, cuttlefish DNA for the ability to change its color, etc. From what I could tell from the movie, the scientists seemingly skirted the design-build-test process entirely and used a computer program to create *Indominus* on the first try. Meanwhile, my lab and I can't even engineer a much, much simpler bacterium to perform one task without repeated iterations. The movie's writers are clearly downplaying—or simply ignorant of—the complexity of biological life.

For the protein degradation project, I created genetically modified bacteria through the introduction of plasmids. **Plasmids** are a piece of circular DNA that I designed, literally to the letter—specifically a combination of A, G, C, and T—that was shorthand for the DNA bases. I treated the receiving bacteria so that their outer membrane would permit entry of foreign material. Then, I would often shock, chill, or chemically treat my bacteria in order to transform them with my plasmid. During transformation, my designed plasmid would hopefully diffuse into the bacterium and incorporate itself into the cellular machinery. In my project, the plasmid would remain within the bacterium but not necessarily change the native DNA of the organism, though it can if I designed it to. If everything went correctly, the bacteria started exhibiting new behaviors conferred by the plasmid DNA. These behaviors could range from, perhaps, glowing green when sensing poisonous arsenic,[135] producing a biodegradable plastic,[136] to making life-saving insulin,[137] all established applications of the genetic engineering of bacteria using plasmids.

In these applications, designing the DNA is only the first step toward the desired function; there are many intermediate steps. It starts with a gene, a continuous sequence of DNA, being transcribed into a complementary molecule, or RNA. Likewise, RNA exhibits defined, coded sequences. RNA is then translated into protein by cellular machines, ribosomes. Ribosomes clamp and slide across RNA. Imagine a cassette tape is the RNA and the tape player is the ribosome; together they generate the music you hear. As the player runs across the tape, it constructs the physical protein in real-time. Each note is a different amino acid, and the entire song is the protein. Once the protein is completed and released from the ribosome, the function is finally available: a group of proteins in yeast can turn the sugar into ethanol; a protein can combine with iron to transport oxygen in your blood; proteins read DNA to make the complementary RNA sequence; or a protein can give cheese particular properties (meltability, stretchiness). When synthetic biologists design DNA, they sincerely hope for a desired function. Given the complexity of the conversion process of turning DNA to RNA to protein, however, many detours occur, exemplified by my pursuit of targeted protein degradation.

The number of proteins and molecules within life is staggering, unknown, and complex. We have successfully genome-sequenced humans, i.e., have mapped their full DNA sequence. In the doing, we learned that humans have in the neighborhood of 20 thousand protein-coding genes.[138] Genes are the DNA sequence encoding the eventual protein, but proteins can be further modified after synthesis, leading to variants with different functions.[139] Therefore, 20 thousand genes can lead to many more different proteins. The HUPO Human Proteome, an international protein research consortium, seeks to ascribe function to at least one protein from each of these genes. While the scientists have made tremendous progress and inferences, still about ten percent of proteins (around two thousand) have a completely unknown function.[140] And this number is certainly larger, given all the variants of proteins, known as post-translational modifications.

The unfamiliarity is not limited to just the proteins in a cell. We have not determined the precise or even rough functionality of large swaths of non-protein encoding DNA, nor all the individual metabolite molecules that show up on mass spectrometry measurements. For example, there are at least a thousand biological metabolites in the well-studied *Escherichia coli* to which we have no attributed function.[141] This phenomenon applies to our food, too; ingredient labels do not accurately represent the constituents within the food product. If you buy a banana, the ingredients will be labeled "banana," but a banana itself contains at least hundreds of thousands of different molecules, i.e., glutamate, malate, fumarate, and many more that neither you nor I have heard of.

To complicate matters, DNA, protein, and molecules do not necessarily have to serve one specific function; they can "moonlight" and perform auxiliary functions. For example, an enzyme that primarily catalyzes the formation of molecules may also block cell division until the cell is ready.[142] Finally, even the unknowns are unknown; there still may be more genes and more molecules that we have not detected.[143]

We have so much uncertainty because finding and characterizing molecules, DNA, proteins, etc. in a biological system requires specific techniques and instrumentation. For example, a mass spectrometer can be used to document different molecules and proteins. Mass spectrometry measurements, however, depend on the way the analyte samples are prepared and how the instrument is operated. These parameters may miss specific proteins or molecules. All to say that our ability to investigate biology depends on bounded instrumentation and techniques, which in turn limits what we can learn.

In the aughts, researchers sought to create the simplest living organism by stripping out genes. The J. Craig Venter Institute runs the Minimal Genome Project which has created a "synthetic" bacterium by chemically synthesizing a genome completely and then transferring the synthetic genome into a host bacterium.[144] The host adopted this new DNA as its own, and subsequent progeny were created, encoded only by this synthetic DNA sequence. This effort must be qualified in how "synthetic" it really is, as the researchers did not completely invent a new sequence; their sequence was ninety-nine percent genetically similar to that of *M. mycoides*. So, it's not the *Jurassic World* software program nor the child from *Splice*. They merely wanted to demonstrate the ability to chemically construct a bacterium's whole genome on the path toward creating the simplest lifeform. In a follow-up study, they showcased a more stripped-down version of the organism. They reduced the number of genes from a thousand[145] to about five hundred.[146] They also spent much time and effort to characterize the different genes and the proteins that they encoded. Nonetheless, approximately one-hundred-and-fifty genes/proteins of the five hundred plus have a completely unknown function. We still don't know what thirty percent of the simplest lifeform's genes actually *do*, but they're essential. If we remove any of them, the organism will not live.

Additionally, the path to function is not nearly complete once the protein is made. For instance, the number of proteins can make a difference. In Appendix A, we discuss how, when the FtsZ protein reaches around 2 thousand copies per cell, division commences. However, if this number dropped below that threshold, even to say 19 hundred, the cells would not divide. So, if another protein sufficiently blocked the gene for FtsZ or another protein chewed up

FtsZ, then the cells would not divide. In actuality, all of these different factors interplay and fine-tune the cells for their evolutionary objective.

Likewise, the disease phenylketonuria originates from low levels of just one protein, phenylalanine hydroxylase. This protein enables humans and other lifeforms to consume phenylalanine, an amino acid found abundantly in protein-rich foods like egg whites, chicken, legumes, and nuts. Phenylketonuria sufferers must monitor blood levels for elevated phenylalanine and fastidiously eat phenylalanine-depleted foods, all because these sufferers lack sufficient quantities of one protein.

What happens when a phenylketonuria sufferer consumes too much phenylalanine? The phenylalanine will accumulate and lead to detrimental effects; specifically, toxins will accumulate in the brain, potentially to the point of permanent brain damage, or a new biological state. Biology is born of more than just genes. In the case of the phenylketonuria sufferer, it is the combination of eating too much phenylalanine and the individual's genes that contributes to the potential adverse state. Both genes *and* **environment** matter.

Our bodies demonstrate how both genes and environment dictate our eventual biology: we have the same genetic information in most cells of our body, yet the same DNA can make hair, nerves in our brain, the lining of our intestine, or the bones in our body. The effect that environment has on these cells means that for different organs and different cell types, particular regions of the DNA are transcribed to RNA at different rates, leading to different amounts of protein and leading to vastly different morphologies and functions.

To complicate biological understanding further, the road from DNA to RNA to protein can also meet impasses, short cuts, traffic jams, and even complete detours, leading to a different destination. Many biological mechanisms ultimately affect the protein's function, abundance, or **activity**, i.e., how fast it works, leading to a different outcome for the biology. Here is a list of mechanisms, off the top of my head, meant to overwhelm you. Don't worry about trying to understand all of it:

- transcription factor repression – protein blocking the DNA to RNA conversion.
- transcription factor activation – protein increasing the DNA to RNA conversion.
- DNA histone modification – proteins that wind DNA. The degree of coiling.
- DNA methylation – DNA can be chemically modified by the organism to convert from DNA to RNA at different rates.
- RNAi knockdown – specialized RNA can block parts of RNA and DNA.
- ribosomal binding site variation – the sequence in front of a gene affects how strongly the ribosome binds to the RNA thereby affecting the protein production rate.
- promoter strength – the sequence in front of the gene affects how strongly RNA polymerase, the enzyme that converts RNA from DNA, binds to a sequence of DNA
- operon order – in lower species such as bacteria, the order of genes can affect how quickly they're converted from DNA to RNA

- protein surfactants – proteins that will sequester DNA from access to RNA polymerases by creating a phase separation; think oil and water.
- protein allostery – proteins that can bind to small molecules, creating different activities.
- protein complexes – proteins physically combine to have different activities and/or functions.
- protein degradation – proteins can be destroyed in different contexts.
- controllable protein aggregation – proteins will clump and lose collective activity.
- protein post-translation modification – further chemical modifications to a protein after it's synthesized, potentially changing both activity and function.

The takeaway is that biology is complex. Genetic engineering is not simply stacking the right LEGO pieces to get what we want. We're never going to be able to distill all of these interactions and molecules into a computer program.

As a result, we only have, in the grand scheme of things, rather crude **computer-aided design (CAD)** programs for specific, narrow aspects: designing the DNA sequence, predicting very, very roughly how fast a protein is synthesized,[147] and finding the best DNA assembly strategy.[148] We do not have CAD programs that tell us how to go from DNA to constructing an eventual organism. And highlighting the vast complexity of biology, I'm dubious about the possibility of a full-fledged CAD program ever existing for synthetic biology. As shown in Appendix A, for chaotic systems such as weather, predictability is fundamentally impossible due to our physical reality. Biological engineering may be subject to the same constraint. Therefore, we should not expect the software in *Jurassic World* to be inevitable. Rather, I think it's impossible.

There's still a long way to go in understanding to the fullest the world of biology. Most new biochemistry papers are about biomolecular interactions, where protein X will bind to DNA Y leading to outcome Z. Furthermore, we're still seeing new papers about new biological modi—adding to that long list above. So, as we learn more, biology becomes *more* complicated, even as we've learned that these interactions often serve emergent behavior. For example, biology relies on DNA-RNA-protein binding to impart **feedback**. Often there is a feedback effect from the interaction: a protein will block the synthesis of itself. It will block the DNA such that it effectively caps the total number of copies it can make of itself.

Proteins blocking the synthesis of themselves seems counterproductive: shouldn't biological entities wish to maximize their proliferation? Let's entertain the counterfactual notion that proteins are able to synthesize themselves unfettered, or perhaps they even activate themselves akin to the autocatalysis that bacteria and yeast are capable of. In such a scenario, the body or cell holding the protein will be utterly overwhelmed to the detriment of the entire organism. In the long run, the protein actually harms its own proliferation by saddling the vehicle by which it propagates. While there is more of it locally, there is likely less in the world because the carrying organism bears too much, and the organism itself cannot reproduce as efficiently. Biology routinely employs this **negative regulation** in which a biological entity (e.g., protein, molecule) actively slows its own production. This is similar to how our brain will inhibit signals after some time. For example, we will habituate to a strong smell after enough stimulus.

THE ROBUSTNESS OF BIOLOGY

Negative regulation is a way for a biological system to meet its ultimate goal. Despite the complexity and the teeming rainforest of molecules and interactions within, *it's all to meet evolutionarily imposed objectives*. That is, to grow into a reproductive state, to reproduce, to consume the necessary nutrients to reach such a reproductive state, to know *when* to consume (e.g., by feeling hungry), etc. Therefore, despite the colorful, variegated soup within, all these interactions and regulations accede to higher, emergent imperatives.

Herein lies the big fallacy when thinking about GMOs and biology generally: evolutionary objectives mold DNA, RNA, proteins, and the molecules within, not the other way around. In other words, genes inserted in GMO strawberries may change and mutate depending on how well they support or thwart evolutionary objectives. The fundamental principle *is* evolution and the associated objectives and niche. We fallaciously presume that because DNA holds the biological code that it's the primary orchestrator. No. The ultimate orchestrator is evolution. DNA is mutated over generations of reproduction and selected by imposed evolutionary objectives. DNA is a means to an end.

DNA does not remain constant as the GMO strawberry replicates itself across many generations. As it mutates, winners are selected by how well they satisfy evolutionary objectives. Furthermore, given all of these mutations, biological systems require redundancies and fail-safes. They cannot be dainty flowers that wilt at the first aberrant DNA mutation leading to malfunction or destruction; that would be a poor winner and unlikely to meet the objective long term. As a result, all naturally occurring biological systems have **robustness**. When an adverse environment occurs or a metaphorical leak springs within, the system can manage. For example, just about all known life has mechanisms to fix and correct DNA (though not to one-hundred percent fidelity). If we didn't have such mechanisms, we'd be walking bags of tumors. Likewise, patients with phenylketonuria are not doomed to imminent death. Perhaps aided by moonlighting enzymes and feedback mechanisms[149] (which help limit the amount of phenylalanine),[150] some PKU patients manage to automatically control elevated levels of phenylalanine.[151]

In order to meet evolutionary objectives, the organism's biology will use whatever available means necessary. Consider the variation of a single protein in a species. Suppose the variation causes the protein to perform all sorts of different functions—to stick to another protein to change activity, to bind to various parts of DNA, to adhere to cell wall components, etc. Any of these functions may benefit the cell or not. The benefits will be carried forward and amplified in consequent generations. Biology is agnostic to the *how*, just as long as the job gets done. It's the little kid who, when asked to clean her room, simply slides all of her toys under her bed. Every organism will look for expediency and use MacGyver-like resourcefulness, especially under the burden of the Pareto frontier. This is why moonlighting and complicated mechanisms are so common in biology.

Therefore, biology is only controllable if all of the evolutionary objectives can be controlled as well. My engineered protein-degrading cells would lose their custom machinery if they replicated enough times because the machinery I manipulated into them burdens the cells, keeping them from meeting their evolutionary objectives. For just this reason, I kept a stock of the cells in the freezer, ones that had been transformed with the DNA, and from these I only cultivated a limited number of generations. This in effect also stymies GMO technology because any meaningful change that we want (producing drugs, fighting diseases, building materials) opposes the organism's objective, meaning that these organisms are evolutionarily barred from performing well in a natural environment. For example, Roundup Ready crops are genetically modified to be resistant to the weedkiller Roundup. This trait is not useful ecologically, i.e., outside the context of a farm, and is shown to adversely affect the crop's ability to replicate.[152] Therefore, one wouldn't expect the gene to persist off the farm should it migrate into neighboring crops and exist in perpetuity.

Accordingly, most genetic engineering efforts, thus far, have been modest. Up until this point, only a few genes are generally introduced because we are limited by complexity and robustness. The more features we introduce, the more our system becomes exponentially more difficult to test. If we introduce one gene, then we have one gene to optimize. We may change the region of DNA in front of it to change how frequently RNA is transcribed from it, so that we get it to the right eventual protein levels: too little and our protein is not impactful enough, too much and we're drawing resources away from the rest of the cell. Now we want to add the second gene and have to repeat the same process over again. Furthermore, introduction of the second protein also draws away resources from making the first protein, per the Pareto frontier principle. Again, we have to perform rounds of design-build-test in order to achieve the proper *combination* of both the first and second gene/protein. Now add a third gene, and you get the idea.

The complicated design-build-test cycle has been borne out in actual GMO creation. The famous Golden Rice Project undertaken by professors at The Rockefeller Foundation in 1999 sought to address the problem of vitamin A deficiency, particularly in Asian countries where a great amount of rice is consumed but vitamin A sources are lacking.[153] By adding two genes into rice, Golden Rice was created, aptly named for the wondrous golden hue that the vitamin A precursor, beta-carotene, imparts (the same molecule that helps color sweet potatoes and its namesake, carrots). The first gene product, phytoene synthase, comes from daffodils, and the second one, phytoene desaturase comes from the soil bacterium *Erwinia uredovora*. These metabolic enzymes transform existing compounds (metabolites) within the rice to other compounds. Exactly as the name suggests, phytoene synthase enables Golden Rice to make phytoene from latent metabolites. The phytoene desaturase, also as the name indicates, will desaturate the phytoene, yielding lycopene, a process similar to how our body converts saturated fats into unsaturated fats. Serendipitously, a third gene to chemically convert lycopene to beta-carotene was not needed, as the rice had the necessary enzymes to perform that chemistry out of the box. Importing the genes into the rice was hardly straightforward, however; a variety of shades of yellow were produced from the resulting grains and the most vibrant hues were handpicked by the scientists as dictated by the design-build-test paradigm.

Unfortunately, upon publication in 2000, Golden Rice met with much opprobrium. Critics charged that it was not producing enough vitamin A to be helpful, failing to concede that the research was merely an initial step and that some vitamin A is better than none. Indeed, in 2005, Golden Rice 2.0 came out, producing 23 times more beta-carotene than version 1.0.[154] The researchers hypothesized and confirmed that the phytoene synthase was limiting the overall carotene production within. They tested a series of candidate synthases, with the one taken from corn getting the job.

THE COST OF ANTI-GMOISM

There is likely more potential for Golden Rice and much more optimization to be done (e.g., balancing the levels of enzymes), but public opposition has slowed further development efforts.[155] In particular, organizations such as Greenpeace have bludgeoned the field of research wholesale and resisted efforts to promulgate any kind of GMO agriculture. As of June 16th, 2019, their website reads:

> *What's wrong with genetic engineering (GE)?*
>
> *Genetic engineering enables scientists to create plants, animals and micro-organisms by manipulating genes in a way that does not occur naturally.*
>
> *These genetically modified organisms (GMOs) can spread through nature via cross-pollination from field to field and interbreed with natural organisms, thereby making it impossible to truly control how GE modified crops spread. GMOs cannot be recalled once released into the environment*
>
> *Because of commercial interests, the public is being denied the right to know about GE ingredients in the food chain. It is therefore losing the right to avoid them, despite the presence of labelling laws in certain countries.*
>
> *Biological diversity must be protected and respected as our shared global heritage. Governments are attempting to address the threat of GE with international regulations such as the Biosafety Protocol.*[156]

Greenpeace defines genetic engineering as manipulating genes in a way that doesn't occur naturally. First, this is a clear appeal to the naturalistic fallacy, which we dismantled in Chapter 1 for imprecision and wrongness. Secondly, we've been genetically manipulating crops and animals for *millennia*. In Chapter 1, we saw that the ancestral banana looks like a caricature of the modern one, because it is only through many generations of breeding and crossbreeding that we have reaped a meatier, more delectable progeny. Breeding and crossbreeding *manipulate* the genes of the bananas. This is genetic engineering. To put it simply, every food that we eat today is a GMO per Greenpeace's definition.

Genetic engineering of foods is ultimately a method, a means to develop a new product. Being against its use would be like being against using scalpels because of bad surgeons or being against 3D printers because guns can be made with them. Genetic engineering is agnostic to the outcome: a genetically modified organism may be beneficial, adversarial, or, most of the time, unremarkable. As a result, we cannot collectively paint all GMOs as having monolithic qualities because the *details* of the change they create matters. For example, we cannot draw some claim about the overall healthiness of GMO products. The question is ill-posed. Golden rice will likely be healthier to someone with vitamin A deficiency whereas an almond engineered with extra cyanide would be awful. Each case of a GMO must be evaluated separately.

To be charitable to the sentiment here, I suspect a Greenpeace representative would respond that genetic engineering as a tool is irreversibly dangerous, akin to nuclear weapons, a rogue artificial intelligence, etc. For existential reasons, I anticipate this representative to say, thusly, that we should steer clear. I'm extremely dubious of a potential doomsday scenario with GMOs. As concluded in the last chapter, any engineer or company would be more interested in having *simpler* organisms.

In the food space, we want organisms that can satisfy the nutritional requirements of all sentient beings. In the medicine space, we want GMOs that attack and nullify pathogens; extirpate embedded HIV sequences; deplete phenylalanine in patients with phenylketonuria; or help repair tissue. In the environmental space, we want to use GMO technology to revive prehistoric chestnut trees that can vacuum in the carbon dioxide from the atmosphere in order to swing the tide against climate change.[157]

Even if some maleficent terrorist organization wanted to build an *Indominus rex* or a super pathogen to wipe out a population, they would have to employ the design-build-test paradigm. Suppose they were trying to make a pathogen with the lethality of Ebola, the furtiveness of HIV, and the contagiousness of the common cold. First, they would have to create many variants, then modify the DNA for proteins that are only partially understood, and would then have to test their creations somehow, likely in live humans. It would be—by leaps and bounds—the most difficult synthetic biology concoction man has ever seen. And each cycle would probably take years just to test. Additionally, such miscreants would be fighting against the Pareto frontier of biology. If a pathogen is more contagious, then it's less likely to be furtive or lethal. Each of these traits requires resources within the pathogen and a pathogen would be too lumbering to have all three traits.

Furthermore, I wish to emphasize that, despite my incredulity of potential harm from GMOs and GMO technology, I support regulations and commission boards to promote their safety. Certainly, we want governments and organizations to secure and limit access to pathogens as we've done with the last known vials of smallpox secured in the Center for Disease Control in Atlanta and in Russia. We should indeed mandate that companies that synthesize DNA for commercial purposes not sell pathogenic DNA to just anyone. The analogy here is to enriched uranium, which is not exactly purchasable on Amazon.com per regulations from the Nuclear Regulatory Commission.[158] Likewise, there has been tremendous progress in creating GMOs that are chemically containable, meaning that they can only proliferate in controlled environments.[159] For example, bacteria can be engineered to depend on non-standard amino acids supplemented in their liquid media. Experiments showed that after many generations, these strains could not mutate their way out of this shackle. We should welcome and fund efforts to promote and understand safe and responsible usage of GMOs.

Finally, we must disambiguate between the practices by GMO companies, namely Monsanto, and GMO technology itself. Monsanto is often derided for unsavory business practices, such as charging licensing fees, promoting seeds that only work one season,[160] and ferociously litigating all comers.[161] Given that Monsanto also is the most visible GMO company in the world, it makes for a cartoonish corporate villain. Monsanto's reputation and operations should not blinker opportunities for every other GMO developer. We can distinguish problem-creating business practices from problem-solving efforts, discouraging the former and encouraging the latter.

There is a cost to anti-GMOism in possible lost opportunities. Quite simply, we'll be able to solve more problems with the technology. Yes, there are risks, but they're not overwhelming and are amenable to regulation. In fact, the potential gains from GMO technology may even thwart the doomsday scenarios envisioned by Greenpeace sympathizers. Oh, terrorists revived and unleashed smallpox? No worries, we can instantly create vaccines then inoculate everyone against it, using our rapid-response vaccine system, where the sequences for smallpox epitopes are used to manufacture on-demand antigen, via synthetic biology methods. We produce the antigen in our GMO strains, formulate, and distribute to the entire population within weeks, averting any epidemic. We could even install our on-demand vaccine production systems in municipalities, a medicinal fire hydrant, ready to be opened in an emergency. Physically, we could have the system manufacturing within mere hours of notice and setup.

This dream played out to some degree with the efforts to develop a vaccine for the 2019 Coronavirus as quickly as possible. Traditional vaccine development requires ten to fifteen years.[162] For the 2019 Coronavirus, the first to the finish line were Pfizer and Moderna, who were able to accomplish this feat thanks to RNA vaccine technology,[163] as opposed to traditional protein, or antigen-based, vaccines. In RNA vaccines, a synthetic RNA sequence is made, i.e., genetically engineered, that our bodies are able to decode and then mobilize our immune system against a pathogen. Moderna's technology is highly adaptable and fast; they created their candidate coronavirus within two days of the virus sequence being available, January 13, 2020 to be precise.[164] Furthermore, RNA vaccines are much faster and scalable to produce. Traditional flu vaccine production can take months because each vaccine is grown in a chicken egg.[165] We've seen just how plodding and inefficient animal technology is; thus manufacturing the seasonal flu vaccine in the United States requires 150 million eggs per year over 2-4 months (for roughly 200 million doses).[166] In contrast, with GMO-based RNA vaccine production, a bench-top scale, two-liter bioreactor can produce one million vaccine doses per run of about a week.[167] So with the equivalent of one small brewery-scale reactor, we have enough vaccine in a week for the entire United States.

To further allay GMO wariness, it's easier to synthetically build biological defense versus biological offense, such as weaponizing pathogens. Consider what it takes to build an offense: the precarious, unpredictable concert of protein, DNA, and membrane are necessary to form a pathogen. The pathogen would also need functions to invade the host, evade the immune system, and hijack machinery. That's a lot, and the Pareto frontier tells us that there's no free lunch: it becomes exponentially more difficult and unlikely with each needed function. In contrast, ramping up biological defense requires less complexity. A vaccine is usually a simple mix of RNA, protein (antigen), or viral particles; it's so much easier to design and produce than a pathogen. Given the tractability difference, I predict synthetic biology research will potentiate the defense much faster than the offense. And this tractability difference doesn't just apply to infectious diseases, but also to the shift away from animal products.

GMOS AND ANIMAL TECHNOLOGY

In order to build genetically modified organisms, we require facile, effective genetic-engineering tools to manipulate the DNA. Scientists have many tools available when modifying the DNA of bacteria specifically, many with great efficiency.[168] Even if a tool doesn't have great efficiency, we can "select" for the bacteria with a genetic change. For example, suppose our bacteria need the amino acid leucine in order to live. We supply leucine in the nutrient media broth to enable the bacteria to grow and duplicate. Now I wish to introduce (transform) my protein-degradation system into the bacteria. We can bundle the protein-degradation DNA with the gene that allowed the bacteria to skip the leucine requirement. Then we introduce the DNA and grow the bacteria in the media without leucine. Even with poor DNA transformation efficiency—say only 0.01% of our bacteria receive the new DNA—the ones with the "don't need the leucine" DNA will be able to grow in the leucine-less media. This selection process can be applied to any microorganism that can grow in bioreactor conditions.

We generally lack such tools and selection techniques for animals, so higher transformation efficiency is required. I've mostly seen efforts in mice[169] and rats[170] serve as a model for human disease research, but very few in livestock animals.[171] Furthermore, genetically manipulating animals generally requires that we modify their embryos. Genetically modified adults will only pass one-hundred percent of those changes if *all* of their sex cells were also modified with the same manipulations. In order to make GMO animals, we would have to make genetic changes to an embryo, perform *in vitro* fertilization, and carry its development forward into adulthood. These processes are laborious, slow, and can easily go awry. Remember Dolly, the sheep who was the first cloned animal in the mid-nineties? She was created in a similar process. While animal cloning is better than it was twenty-five years ago, it's hardly commonplace, and animal cloning projects have been modest, with roughly about a thousand animals cloned since.[172] Cloning pets still requires substantial costs; for example, a California couple paid $50,000 in 2020 to have their dog cloned.[173]

Finally, we must test our genetic changes to confirm whether they have conferred the functional changes that we're hoping for. The biggest limitation to testing is the time it takes for the organism to grow and reach maturity. For microorganisms, this maturity time is mere hours, as stated in the last chapter. We can start testing bacteria within half a day of genetic change, but it'll take us years before we answer the question for a cow. Furthermore, we have an advantage of possessing a large number of cells and controllable conditions for bacteria. We can grow them in a bioreactor and even control how fast they grow to ensure that what we're testing for is statistically valid.

In sum, the use of genetic engineering in animal technology doesn't measure up to the same use in bacteria and yeast because it is so hard to perform design-build-test cycles with animals. We can innovate with bacteria and yeast faster than we can innovate in animals. In the epistemology from Chapter 2, *bacteria and yeast are much more tractable than animals.* While genetic engineering helps the tractability of animal technology, it helps bacteria and yeast vastly more so. Therefore, the availability of GMO technology means that we're less inclined toward using animal technology.

I beseech animal rights activists to help promote wider acceptance of GMOs. I suspect a significant fraction of animal rights activists, vegans, flexitarians, and vegetarians regard the genetic engineering of foods poorly.[174] This *is* a tradeoff because animal rights and opposition to GMOs directly clash. Genetic engineering will help catalyze the end of animal-based consumer products. I suspect that the mindset is ultimately due to a lack of knowledge. Therefore, if you're struggling with an anti-GMO stance, I implore you to further investigate what genetic engineering entails. Learn more about the central dogma in biology, participate in a community lab, watch videos on how CRISPR-Cas9 works, and talk to scientists. Finally, I can't emphasize the limits of engineering biology enough. These physical limits we've discussed above mean that it's much easier to do good with genetic engineering technology than evil.

THE FUTURE OF GMOS

In the last chapter, we discussed how the fundamental physics of biological production pushes us toward using smaller organisms. Small organisms also tend to be more tractable for modification. However, we still struggle to control evolutionary objectives. Species will always veer toward maximizing their optimization function, generally dispensing with whatever modifications we make unless we continually apply our own selection (e.g., choosing which crops to cultivate) or reseeding the process (like my frozen

DNA sequences that I used periodically to refresh my ongoing production).

To these points, I see the current modus for biological production staying in use for a while: make modifications to chassis organisms, such as yeast, in order to produce a desired product; and freeze and store a yeast stock that's only been replicated a limited number of generations. The yeast stock is similar to a sourdough culture, with far more control and validation behind it. We can use this yeast stock to seed a fermentation process, harvest, and separate our product. In order to restart production, we'd simply return to the yeast stock. Never should we seed a fresh bioreactor using yeast from a spent process because that would only add more generations. By only starting with frozen ones, we limit the number of generations for each individual reaction. This problem currently rules out a continuous process; we cannot just run a bioreactor without stoppage. However, the process metrics of microbes are so incredible that they still overwhelm an animal-based process.

When tractable enough, we will use complete designer organisms whose evolutionary objectives we can more precisely impose. This xenobiology could also be optimized for a bioreactor environment, leading to even better productivity and yield metrics. We could conceivably even develop a continuous process. Until then, I see us engineering naturally occurring microbes to produce target substrates, as we have done to harvest insulin, heme, and alcohol. In general, the one substrate per bioreactor operation should continue to pay dividends for a while, and there is still so much room for development, especially once the shackles of anti-GMO hysteria have been broken and research is freed of artificial barriers related to public opinion.

To illustrate my point, consider egg whites, which are used in sauces, dressings, and desserts. The majority constituent of egg whites is albumin, a protein known for its ability to dissolve well in water. Albumin denatures when subjected to heat: the arrangement of amino acid chains within the protein fluctuate and knot in ways that destabilize the entire product. This misfolded protein state is insoluble in water (think oil) and separates out. As the heat diffuses through, the albumin in the solution misfolds, and the pariahs band together into a solid mass. Therefore, fluid with albumin can be scrambled, formed into an omelet, and used to bind ingredients in cakes. The process is even reversible. A surfactant can be used to help resolubilize eggs such that, when added, the cooked morsels dissolve back into the more liquid, slimy, thick egg white.

As discussed earlier, denaturation works for any soluble protein to varying degrees, and vegan egg replacers have adequately accomplished the task. Potato starch, soy protein, and lentil protein have all been enlisted in these endeavors. Therefore, why don't we produce such a protein in a bioreactor? It would be dirt cheap, fast, and obliterate chicken eggs in terms of techno-economics. The product JUST™ egg relies on an expensive, poor-yield separation process to extract the protein from lentils. The inefficiency is reflected in the price, which certainly deters thrifty consumers and ultimately forestalls full-on replacement. I see a potential bioreactor process for a GMO-albumin analog already tractable to develop with current knowledge and technology. I surmise that egg whites would be replaced if everyone were open to it. In fact, the startup Clara Foods is working on producing egg albumin in a yeast fermentation.[175] Clara's efforts would be far easier if they could develop an albumin analog instead of feeling compelled to produce egg albumin to molecular exactness. Note: we will discuss the taste aspect in Chapter 8.

Once the single-product use cases take root, we can look at multi-substrate applications. In Chapter 2, I imagined a personal

3D printer that could fashion an on-demand steak at a moment's notice. Such a printer would need ink (proteins, fat, and other biomolecules). The "ink" would undoubtedly be supplied by bioreactor production. We could have three different kinds of protein. Maybe one of them is a scaffold that structurally holds the steak elements together in recognizable form; another provides the striated texture; and the third imparts a delectable umami flavor. Our semi-solid fat ink marbles into a protein matrix, and various biomolecules optimize the taste and nutrition.

Speaking of nutrition, I suspect this concern—that engineered food is somehow inadequate or lacking in vital nutrients—might also be holding humanity back from making the full leap into a future without animal products, so let's go there next.

CHAPTER TERMS

- **genetically modified organism** (GMO): an organism whose DNA was changed by intentional means. All foods we eat today have had their DNA intentionally modified; all foods we eat are GMOs.
- **plasmid:** a circular piece of DNA that can be synthesized to the letter and introduced into microbes
- **design-build-test** : an iterative approach to engineering a solution where researchers or engineers pursue each step in sequence, repeating the process with refinements once they observe the results; this is how all experimental engineering efforts are pursued
- **computer-aided design (CAD):** a engineering pursuit facilitated by a computer program that abstracts the underlying principles. The software performs all calculations in order to achieve the

desired outcome. All CAD programs have a limit in precision and scope.

- **environment (biology)**: the surroundings of a biological organism, which influences development as well as evolution
- **(protein) activity**: an enzyme may have different catalysis speeds depending on chemical modification, binding from other molecules, interactions with the solvent, etc. Activity is the amount of catalysis a protein can perform for a certain duration in its given context.
- **feedback**: changing the rate of an upstream process by how a downstream process occurs. For example: if I don't sleep well at night, then perhaps that's feedback to lay off coffee after 12pm.
- **negative regulation**: feedback that specifically curtails the upstream process when the downstream is too abundant
- **robustness**: the persistence of biological systems to meet their evolutionarily imposed objectives even after vicissitudes, such as DNA becoming mutated, becoming injured, getting engineered, etc.

CHAPTER SUMMARY

Lack of knowledge, inherent complexity, susceptibility to feedback, environmental effects, and robustness to evolutionary objectives—all of these factors render biology arduous and limited when it comes to engineering. A full CAD software for biology might be fundamentally unobtainable with native systems, whether bacteria, yeast, plants, or animals, and the design-build-test cycle modus remains the foreseeable standard. The lack of knowledge limits the scope of predictability. The feedback subjects our design to countervailing forces. The complexity necessitates a ballooning number of equations as we seek more precision. The environmental effects mean that we often need to sequester our biological system into controllable conditions (e.g., a bioreactor) in order to maintain the desired functions. And the robustness of biology means that we'll generally face resistance if opposing the objectives from the species' initial evolution. These features are more limiting and disabling in larger, slower-growing organisms. Using genetic engineering helps overcome these challenges for engineering purposes, but it's more effective where we have more knowledge, less complexity, and rapid prototyping. Therefore, in a society using GMO technology more acceptingly, we'll see animals displaced faster. The goals of animal-rights activism generally collide with anti-GMO ones, most of which are promulgated through naturalistic impulses, poor understanding, and insufficient argumentation. Future cataclysms from GMOs are unlikely, especially with intelligent oversight, and the technology will only ennoble our food production.

PART THREE

Replacing Animal Products in Food

CHAPTER SIX

Nutrition and Animal Products

THE DIFFICULTIES OF NUTRITIONAL SCIENCE

Earlier we discussed the four Ns that accompany and support eating animal meat (natural, necessary, nice, and normal), focusing on the first term. Now, we will tackle and refute the next one: necessary. Necessary implies that meat is nutritionally essential. Humanity spends a lot of time, money, and sanity on nutritional science.[176] But rigorous nutritional study is difficult; for example, how would we know whether bananas are truly good for everyone? For argument's sake, let's do a thought experiment. In this experiment, we need a control, non-banana-eating group and an experimental, banana-eating group. Ideally, the two groups are exactly the same except for banana consumption: clones with the exact same experience leading up to that point because both genes and environment matter, as we discussed in previous chapters. Both groups must be fed the exact same diet and have the same daily activity. Let's just have them in prison and control their routines to the minute.

Furthermore, we cannot simply remove bananas from the control group's diet because what if the differences between the two groups is simply explained by the calories that the bananas provide. We must substitute for the bananas in the control group. So, what's the correct alternative? Apples? Sugar? Wafer crackers? Additionally, our study is longitudinal, so we must keep the participants in prison for years. We take blood and document vitals. We also assess for strength with evaluative challenges. And we finally conclude whether bananas are better for us than apples, wafer crackers, or nothing.

Obviously, this is absurd, unscalable, near impossible, and immoral. These traditional, rigorous experimental strategies used in the domains of molecular biology and physics don't apply well to nutritional scientific pursuits. Instead, **epidemiological**, longitudinal studies are employed, where two similar populations are chosen except for the experimental condition (e.g. one smokes). These populations are then tracked for specific outcomes. For instance, a higher incidence of lung cancer may be associated with the population that smokes, leading to the hypothesis that smoking causes lung cancer. Of course, other effects need to be accounted for, too. Perhaps coal miners smoke way more than everyone else, and they made up some part of the subject cohorts. But in reality, not all coal miners are smokers, and not all smokers are coal miners, so epidemiologists can separate the groups and causes of lung cancer. And when the cohort enlarges, i.e., more people are studied, we can be more confident in the conclusions.

The epidemiological method works best when the effects are pronounced. Capturing the problems of smoking, asbestos, lead paint, and vitamin deficiency are easier than finding out the effects of bananas, coffee, chocolate, or red wine on human health. In the case of the former, for instance, lung cancer, lung disease like mesothelioma, low IQ, and physical deformities are such pronounced effects that they are easily revealed by an epidemiological study.

The effects food has on individuals are more subtle, particularly because our bodies are not hard clay that gets chipped away or added to by particular foods and habits. The dynamic feedback and robustness discussed in the last chapter mean that human bodies will adjust to what we eat. Therefore, the individual side effects of consuming most food items ultimately will be dampened and harder to tease out from everything else going on inside our bodies. The tools of nutritional science are inadequate to untie the Gordian knot. Furthermore, the questions are often poorly conceived or simply non-computable: why should we expect bananas, coffee, or chocolate to be categorically bad or good for us? Compared to what exactly?

This fruitless pursuit endures because we want pithy directives that guide us to live a long, healthy life *and* gain and maintain washboard abs. Additionally, we want to feel validated while indulging in our vices; i.e., it's okay to have that glass of red wine or to butter our toast. I suspect the truth is we actually know quite a bit about nutrition, but unfortunately it doesn't lend itself well to tidy aphorisms—"gluten bad," "kale good." In particular, we know a lot about metabolism that informs nutritional science. I will not prescribe my version of "Eat food. Not too much. Mostly plants."[177] Instead, I can offer the most value by sidestepping the ubiquitous, but relatively inconclusive nutritional research and use knowledge from other fields (e.g. systems biology, metabolomics) to add another approach to nutrition. As discussed throughout this book, the highest knowledge isn't a scientific study that crunches numbers. The highest knowledge is a stiff, falsifiable explanation (see Chapter 2).

Proponents of eating animals will often cite nutrition as the primary reason to justify this practice: animals are an abundant, rich source of the essential building blocks of life: protein, omega fatty acids, vitamin B_{12}, iron, and zinc. Vaclav Smil answers the eponymous question in *Should We Eat Meat?* by concluding that some meat in the human diet is essential, but most of the developed world eats too much and can cut back. Smil claims the following:

> *"Meat's importance in human diets is primarily due to the supply of high-quality protein, secondarily to the provision of fatty acids and micronutrients and finally as a source of food energy."*[178]

Smil is dubious of cultured meat technology, both in terms of the tractability as well as the acceptance from the wider population. He's also wary of replacing proteins with classic vegan substitutes (e.g., soy, nuts, legumes) because the protein is apparently "lower quality." And it's not clear what that entirely means. To my mind, Smil is discounting just how much potential technological possibility we have to replace the nutrition from animal products and relying too much on imprecise nutritional studies. So, let's first inject more precision into nutritional research based on first principles.

THE FUNGIBILITY OF METABOLISM

When we consume food, we chew it with our teeth, mixing it with saliva from our mouth, to reduce it to a swallowable slurry. The saliva houses specialized enzymes that break down ingredients within our food. In particular, the amylases break down starches into smaller carbohydrate (sugar) molecules. Likewise, saliva lipases do the same with fats. The saliva-mixed slurry will travel down through the throat and esophagus and then into the stomach. There, the fat further breaks apart into free fatty acids. Eventually the fatty acids will be absorbed with the help of bile salts in the small intestine. Protein is hydrolyzed (chopped up) by the stomach acid into its alphabet-named molecules, otherwise known as amino acids. The human body absorbs, circulates, and metabolizes these bite-size components of carbohydrates, amino acids, and fats. This is why we all carry a bag of acid—our stomachs—to help **digest** bigger macromolecules into metabolizable, constituent molecules.

Once our bodies have transformed food into individual molecules, our **metabolism** takes up the baton. For example, glutamate and glutamine are among the twenty amino acids we liberate from consumed protein. These amino acids differ by one chemical feature, an NH_2 amine group (**Figure 12**). If the body needs more glutamate—for example, during low energy[179]—then the glutaminase enzyme can be activated. This enzyme removes the amine group from glutamine, thereby forming glutamate and ammonia (the freed amine group). The reaction can even run in reverse via a separate enzyme (glutamine synthase), which is especially helpful in situations when the body needs to sponge up excess ammonia and curtail brain toxification.[180]

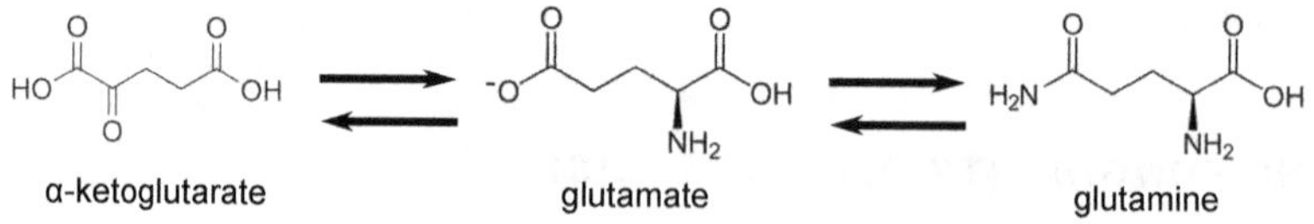

α-ketoglutarate ⇌ glutamate ⇌ glutamine

Figure 12. Interconversion in metabolism. Three metabolites that are common in the human body and all known biological organisms. All three are obtainable from food. Glutamate and glutamine are freed when consumed protein is digested. These molecules can be chemically converted from one to another.

Glutamate can also transform into the non-amino acid, α-ketoglutarate.[181] Another name for this omnipresent chemical transformation occurring throughout our body, facilitated by enzymes, is metabolism. Metabolism doesn't end with these three molecules; they're part of the circular metabolic network, the Citric Acid Cycle (**Figure 13**). This cycle exists within simple organisms such as bacteria all the way up to complex organisms, including humans, with only a few dissimilarities—and for good reason. The Citric Acid Cycle provides a lot of energy at a cheap cost: as molecules are chemically converted into the next entity

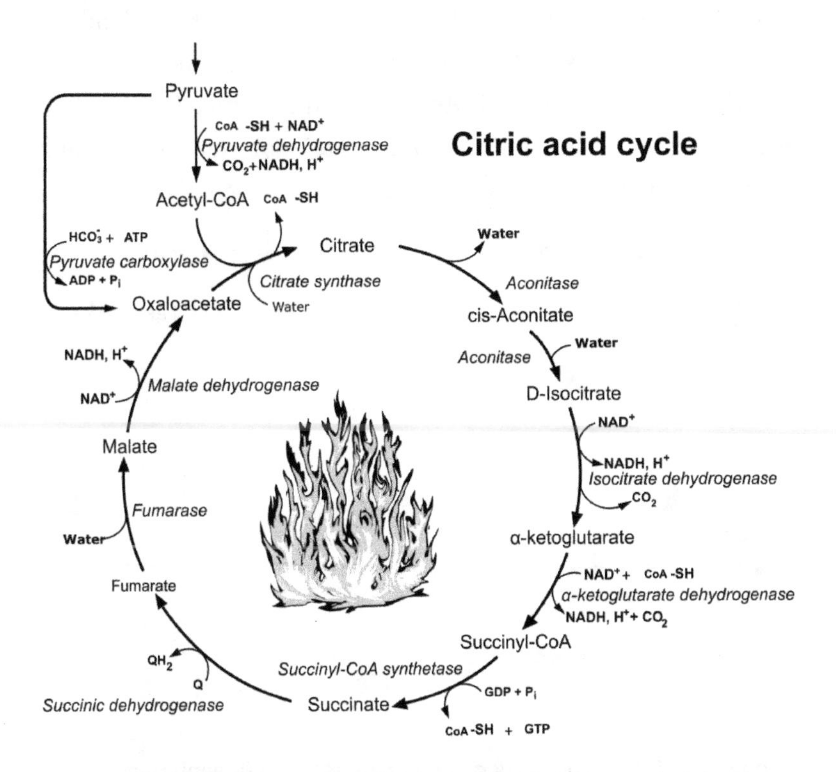

Figure 13. The Citric Acid Cycle is analogous to a fire. Metabolites are converted to the compounds noted around the ring. Along the way, chemical molecules are combusted, releasing carbon dioxide and energy (NADH and QH2 here). This is why we feel warm after strenuous work or exercise, e.g., elevated metabolism.[182]

within the circle, they combust, splitting into carbon dioxide and energy. This energy is usable for everything from pumping the heart to fueling our brains as we read this sentence. The carbon dioxide escapes through our exhalations and through our skin. Fats can be chemically converted and enter the Citric Acid Cycle, too. Therefore, when we exercise, we are actually burning fat, and we lose fat as a gas—carbon dioxide—rather than through excreting it as waste. We ultimately breathe the fat out.

Freeing up energy is not the only task of metabolism; metabolism enables all life to build biomass: protein, DNA, RNA, and lipids (fats). The entire metabolic network is far too intricate and overwhelming to present here; instead, long, serpentine arrows are used to indicate how the Citric Acid Cycle and the metabolic networks above it form into biomass (**Figure 14**). For example, to build protein from sugar, as the bacterium *E. coli* does, each of the twenty

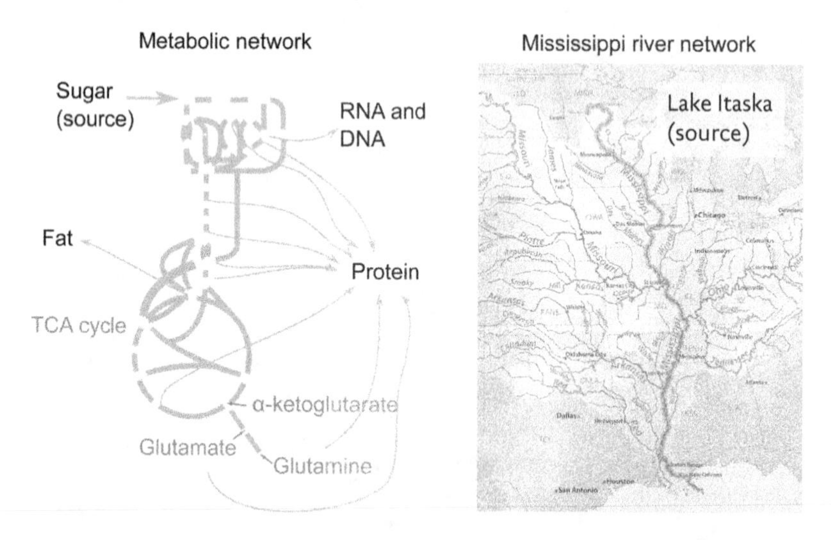

Figure 14. A metabolic network can be imagined as a flowing river. Input of sugar, amino acids, and fat are metabolized (flowed) into energy and the creation of new fat, protein, DNA, and other biomass. Juxtaposed is the Mississippi River, which could be said to have similar attributes of metabolism in that it flows to and from many points.[183]

constituent amino acids must be created, siphoning from different points in the network shown. In my research career, we often described the conversion of molecules in metabolism as a flow, akin to a flowing river branching and ending at different points. The Mississippi River fits as a choice metaphor. The flow starts from Lake Itaska in Minnesota (akin to the sugar in the metabolism example), and the water spreads to other rivers, terminating at many points. And the main body of the Mississippi River ends as it empties into the Gulf of Mexico.

The analogy to a river system is not wholly illustrative. In one aspect, metabolic flow can change directions, as highlighted in the first example with glutamate and glutamine. In another example, if our body needed to create DNA or RNA, we could do so from amino acids, that conversion necessitating an upward flow in the metabolic network. In fact, such **regulation** is evolutionarily developed, allowing us to maintain that biological robustness no matter what we eat. Nonetheless, in most scenarios, we're not just eating protein, carbohydrates, or fats. Generally, we eat some combination thereof, but it all ultimately feeds metabolism, which fuels the Citric Acid Cycle. Furthermore, the Citric Acid Cycle is not picky in source material; it can work with whatever is supplied to it. Similarly, it doesn't matter where it rains for the Mississippi River to keep flowing. Therefore, when the Citric Acid Cycle generates our body's usable energy, the source proteins, fats, and sugars are **fungible**—like the rain in the river analogy—and are interchangeable currency fueling the end result.

To palpably appreciate this fungibility, I witnessed the rapid flow of metabolism with bacteria fed by sugar drops, which is discussed in Appendix A (**Figure 21**). In this instance, we observed spikes of glutamine in the cells within seconds of feeding bacteria the sugar. Similar spiking occurred with hexose-phosphate in the same time span (**Figure 15**). Curiously, we also observed precipitous drops in the metabolites phenylalanine and hypoxanthine. Without going into excruciating detail, the drops of these two metabolites indicate that the cells are making protein, DNA, and RNA. All of this happens faster than each measurement time window (ten seconds). Altogether, the data suggests that it takes less than a second for the flow of metabolism to occur. This observation has been echoed in my other postdoctoral projects as well.[184]

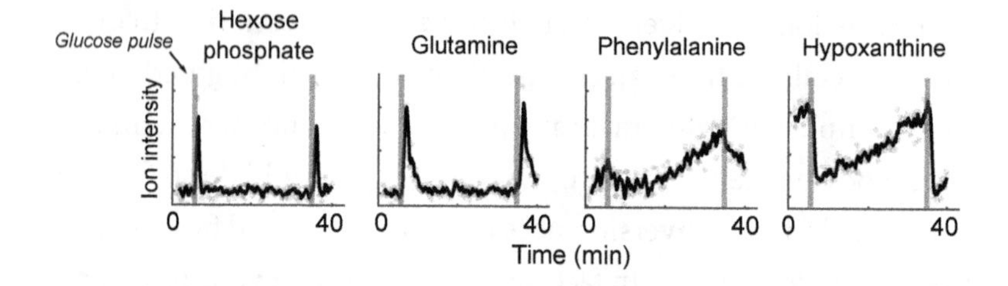

Figure 15. The fungibility of metabolism. When sugar is fed to a group of bacteria, all the compounds of metabolism reverberate in accordance, while pools of the metabolites hexose phosphate and glutamine spike. Existing pools of phenylalanine and hypoxanthine are consumed because of resurgent metabolic activity.[185]

The fungibility characteristic calls into question the true utility of many products on the market. For example, the collagen industry achieves billions of dollars of sales per year, through direct sales of the protein now featured in so many cosmetics, nutritional powders, and foods.[186] Collagen protein is a ubiquitous and central element to our skin. In animal industries, collagen is essentially a waste product, otherwise called ground up and acid-treated bone and cartilage. With some creative marketing, backed by poorly-controlled studies,[187] producers were able to commoditize offal into costly cosmetics.[188] The aspirational message is that for skin as nice and creamy as Jennifer Anniston's, one needs more collagen, which is possible if one consumes some directly. In the past few years, the collagen craze has made its way into food,[189] and now collagen food products are close to a quarter billion dollar per year market. But we don't get collagen into our skin simply by eating it; that does not square with what we've just been discussing.

Rather, the collagen must pass through our stomachs, where it is digested into the constituent amino acids—glutamine, glutamate, etc. These amino acids then fuel metabolism, probably some of that making collagen for skin; but again, given the fungibility of metabolism, it doesn't have to come from directly consuming collagen—it could come from broccoli, soy sauce, or beer. There is no way for us to incorporate collagen in our diet in any direct way that it transmigrates intact into our skin cells. (I challenge a study to show this. Feel free to reach out to me for technical help.) Ultimately, collagen food-marketers and producers play to ignorance about human biochemistry, and I, for one, am looking forward to the demise of collagen-based products and all the foolishness associated with them as either "natural" or "good for your skin."

I should note that not all of metabolism is fungible; some specific molecules cannot be replaced or exchanged. Vitamins and minerals (e.g., zinc and iron) fall into this category. We lack the metabolic networks to make essential vitamins such as vitamin C, and we likewise tend not to break them down. Interestingly, most animals, including dogs, cats, and lemurs, are able to synthesize vitamin C from other elements of their diets; however, we humans can only obtain vitamin C through consumption.[190] Furthermore, there are nine amino acids that we cannot directly synthesize metabolically from fats, sugars, or other amino acids.[191] The famous "turkey" amino acid, tryptophan, falls into this essential amino-acid category, even though tryptophan is found everywhere and not just in turkey. These essential amino acids are recouped from consuming them, or protein, or from recycling the proteins within our bodies. Nonetheless, a large majority of our food intake is fungible or has a degree of metabolic fungibility. This means that we don't need animal products because we can obtain nutrients from a variety of sources.

Even if our cells lack the capability to synthesize some amino acids, our gut microbiota friends can pick up the slack. These microorganisms inhabit primarily our large intestine and make up a significant mass of our stool. Our microbiota adjust to our diets such that meat- and dairy- focused diets, for example, promote microbes tolerant of the bile acids needed to absorb the dietary fat. Diets replete with leafy greens evolutionarily select for microbial Firmicutes that help break down plant fibers and liberate the nutrients for our own body to recoup.[192] Gut microbes can even synthesize some of the amino acids that we can't, thus freeing them for use in our bodies.

We are the end result of dietary evolution over millions of years during which the biggest limitation to survival was just getting enough food. Therefore, the evolutionary objectives shaped our bodies to value every nutrient with the utmost importance. That's why, when we consume too much food, we're liable to make storage-nutrient molecules such as glycogen and fat. Glycogen and fat can supply energy and nutrients in periods of starvation. These features have turned from a benefit to a liability in our current society where food is so abundant.

The body's storage systems are activated when our metabolism is overfed. Specifically, when the sources that fuel our metabolism pools are plentiful—think the Mississippi River overflowing—then insulin regulation kicks in. The storage is, in effect, over-spillage from the metabolic flow. Therefore obesity and fat synthesis will be curbed by curtailing periods of high metabolic pools (partly induced by elevated insulin) combined with low energy demand. Since the Citric Acid Cycle pool fuels both energy and fatty acid synthesis, both actions must be used or stored. To put it plainly: if I have a couple of beers with no exercise, those calories will probably go straight to my belly, stored as fat.

It's difficult to square the fungibility of metabolism with the supremacy of protein. It's not just the keto and Atkin's diet advocates; venerated institutions such as the American Heart Association and American Cancer Society recommend high protein, low carbohydrate diets.[193] And of course, high protein, low carbohydrate diets suggest increased meat consumption. But ultimately, both protein and carbohydrates will indistinguishably fill our metabolic pools, so why should we weigh one over the other? Indeed, there's a deep reason and a good explanation, but first we have to answer whether there is something special about the *amount* of protein in a diet.

THE PROTEIN IMPERATIVE

We descended ancestrally from primates and are appropriately still in the same species family. Species-wise, chimpanzees are our cousins, and we have about ninety-five percent of the same DNA.[194] Aside from body hair, we differentiate ourselves from primates through our diet: we can consume omnivorously, eating both meat and plants—whereas our hirsute cousins remain herbivores. This distinction is the base of a popular hypothesis as to how humans came to dominate the planet: the predecessor to *Homo sapien* was herbivorous, but rapidly evolved to be able to consume animals. In these prehistoric times, this was an incredible advantage because starvation limited species proliferation, and suddenly our species had access to a whole new food source. Another theory suggests that the access to vastly available protein in the form of animal meat enabled our large brains, and with such large brains we could think ourselves into dominance and more complex societal development.[195]

But again, we have to consider the fungibility of metabolism. Couldn't an alternative explanation be that access to more *calories* was enabling? By making the transition from herbivore to omnivore, we had more food, period. We no longer were limited to just plants. We could consume both plants *and* animals. Today, this is no longer an advantage as we discussed in Chapter 1. We have plenty of food and calories. Yes, distribution could still be better, but humanity has no shortage of available calories. Even if animal products had more calories or caloric density, that's not adding much value.

A brief, important aside: eating animals is sometimes excused by the "fact"—itself oft debated—that animal protein allowed the development of our large brains.[196] This is an awful argument. Plenty of immoral, historical actions have sparked some positive outcomes and developments—Nazi medical experiments,[197] British colonization of India,[198] slavery in America, and selling tobacco. We could experiment on humans to improve the health of most of humanity. However, collectively and laudably, we have decided that doing so is not worth the cost of admission. Anyway, we'll discuss the ethics of using animals for products more in the final chapter, but I didn't want to brush this aside while topical.

To distinguish whether the evolutionary benefits are attributable to protein or calories, let's do some math. We can figure out how much protein the human body actually requires. Here, nutritional studies generally fail us, partly because vested interests want to continue the narrative about the "need" of protein, and partly due to the impossibility of performing controls. Instead, we can do the math using a physical, mass-balance approach by which we tally how much protein we use per day to build new blood cells, hair, skin, mass, etc. By the Law of Mass Conservation, the amount of protein that we eat must meet or exceed this value. "Protein In" must be greater than the "Protein Out."

The "Protein In," as we just saw, can become or substitute for fat and sugars, and so, to simplify analysis, *let's ignore the fungibility characteristic of metabolism*. Therefore, what we calculate as Protein In must be the real deal, as if our bodies can only produce protein from protein or amino acids we eat. Now we must characterize and tabulate Protein Out, or the irretrievable sinks of protein we lose per day in order to grow (make biomass); the cells lost everyday (skin, hair, and epithelial cells that line our intestine); and those we turn over in our bodies, such as blood cells (erythrocytes), cells that line our stomach, our skeletal muscles during strenuous exercise, and immune cells. The creation of new cells and tissue requires protein; *however*, protein is undoubtedly recycled from dead cells and tissue as well.[199] Evolutionarily, we could not afford to simply excrete out valuable nutrients. Ultimately, we want to calculate the mass of protein that we cannot recycle as material that has too much molecular damage or because of sheer inefficiencies. This is what must be replenished by our diet.

To simplify further, let's only consider adults who do not grow, leaving just two components of Protein Out: (1) the skin, hair, and epithelial cells we lose for good, and (2) what we can't recycle in our body. We can calculate the first component fairly easily by knowing how fast hair grows and how many skin cells we shed per day, but the second is not nearly as approachable. Instead, we must be more creative and consider that protein distinguishes itself chemically by the presence of nitrogen. Protein contains nitrogen whereas fat and carbohydrates do not. When we consume too much protein or nitrogen, we either urinate it out or defecate it with our dead gut bacteria.

Therefore, the content of nitrogen in the feces and urine is the response variable. If I eat more protein, my excrement contains more. And this should work linearly. If I eat less protein, less is vacated. If I eat the absolute minimum protein required, then I lose the equivalent nitrogen, or the **obligatory nitrogen loss**. If I eat *less than* the minimum, I still lose the obligatory nitrogen loss, but now I run a deficit, and my body will deplete nitrogen over time. Therefore, the obligatory nitrogen loss can be understood as what we minimally require that must be replaced through consumption in our diet in order to be healthy. This value has been calculated to be fifty milligrams of nitrogen per day per kilogram (2.2 lb.) of body weight.[200] As of writing this, I weigh about 80 kg (175 lb.). That means I minimally require about four grams of nitrogen per day. If I'm getting that purely from protein, it amounts to about 25 grams of protein per day.[201] A picayune amount. Even if I ate nothing but white rice (2 thousand calories worth), I'd hit 40 grams of protein,[202] easily surpassing my protein mandate. This is equally true for diets consisting of nothing but vegetables and fruits. It would actually be nearly impossible to *not* hit the protein requirement on a 2 thousand–calorie-per-day diet. I would basically have to subsist entirely on refined, zero-nitrogen sugar, which is often used illogically and misleadingly as the control for industry-sponsored studies on protein consumption.[203]

You might be wondering what the role of exercise plays here, and whether someone needs drastically more protein if she is running or biking. We know that of the 25 grams I need per day, roughly thirty percent or 8 grams are for skeletal tissue.[204] Exercising at a moderate level, eighty percent of maximum heart rate for forty-five minutes, three to four times a week, increases skeletal muscle synthesis by about twenty percent.[205] That's an addition of roughly 2 grams, and still only brings the total protein mandate to 27 grams. Even if I run a marathon every day, extrapolating it further doesn't suggest that much more protein demand. Even if it doubles skeletal muscle synthesis—8 grams to 16 grams—that still only means 33 grams of protein are needed per day.

The situation is different for nursing mothers, growing kids, and babies, who require an additional sink of protein to form more biomass as their bodies enlarge, or in the case of nursing mothers, create milk to help their babies grow. Like microbes, human biomass is mostly protein.[206] Ignoring water weight, we are almost fifty percent protein. Therefore, let's assume for each gram of biomass that we create, we require 0.5 grams (a tiny fraction of an ounce) of protein. Babies, when they grow their fastest, add roughly 25 grams (nearly an ounce) of biomass per day (1.7 lb. per month).[207] This means that babies need about 12.5 grams (.44 oz) of protein per day for just biomass creation. Combined with the amount required for the other sinks (about 1.5 grams or .05 oz), we get to roughly 15 grams (.52 oz) per day. This seems to jive somewhat with current recommendations of 10 grams (.35 oz) of protein per day for babies.[208]

One of the more authoritative recommendations for the amount of protein per day comes from the National Institute of Health.[209] The recommendations account for the nitrogen balance analysis laid out here. Given that these recommendations are applied to everyone, the NIH must be careful. They specifically account for the differences among people and must buffer against this, so they accordingly recommend a bit more than what's calculated to be sufficient.[210] As a result, I'm recommended to consume 56 grams per day of protein, roughly more than double of what I calculated.

Nonetheless, these recommendations are easily satisfied with any diet. Even if I ate nothing but potatoes, I'd hit 56 grams (1.9 oz) per day. If anything, the average American male diet of 90 grams (3.2 oz) of protein per day is vast overkill.[211] Furthermore, we can satisfy the nitrogen balance by consuming amino acids directly (which ooze out of vegetables and plants) or by just consuming nitrogen. We can form proteins from sugar and nitrogen as discussed earlier. So, all of the presented calculations for protein needs are overestimated.

All in all, a high *quantity* of protein in our diet is not something to be concerned about. If you eat only junk food, you'll get enough protein. Therefore, we should not worry nor be intentional about obtaining enough protein, no matter the source. I personally adhere to a vegan diet. I don't bother to count protein and many of my meals will lack a "protein source" (e.g., coconut curry with only vegetables). But, at the time of writing, I have a healthy body mass index (80 kg/175 lb. at 175 cm/5'9" height) and a lower body fat percentage (about ten percent) compared to both my meat-eating and cheese-eating days.

Cases of protein deficiency stem mainly from growing children who may need more protein than a diet supplied by, say, just potatoes.[212] Finding such cases for adults has proven difficult. (Please alert me if you find them.) Health problems attributed to lack of protein are difficult to adduce from lack of calories, and this point extends to the development of human brains. However, I do see a potential benefit in protein related to how slowly it is metabolized. That is worth discussing further.

SLOW BURN AND THE DIGESTIBILITY OF FOOD

So far in this chapter, we're moving away from the "component"-centric view of nutrition and metabolism—obsessing that we obtain 80 grams (2.8 oz) of protein per day or avoiding over 25 grams (.88 oz) of sugar. Our bodies don't treat these ingredients completely separately, so neither should our nutritional directives. We need a different way to approach what constitutes a healthy diet, though healthy can mean different goals for different people and situations. To narrow the discussion, I focus on diminishing obesity while still obtaining all essential nutrients. Most of us interested in nutrition and health, at least, share these objectives.

Earlier, we discussed metabolism as a flowing river, and that the synthesis of fat occurs when it overflows. Maintaining a steady level in the river and minimizing the overflow should be useful to curb obesity. In fact, Professors Eran Segal and Eran Elinav of the Weizmann Institute of Science have incorporated the notion of metabolic overflow into The Personalized Nutrition Project.[213] Participants wear glucometers that track the level of their blood sugar (glucose) over the course of a week. Given the fungibility of metabolism, this measurement serves as a reasonable surrogate to the level of metabolic activity, or in our metaphor, the water level of the river.

In a representative study, the labs of Segal and Elinav sought to reduce the blood glucose spikes that occur after a person eats a meal.[214] However, each participant ate different diets, had varying lifestyles, gut microbiomes, and exercise regimens, and represented multiple demographics. So, the study's authors devised an algorithm to propose a new diet that calibrated the foods and the timing of calorie consumption in order to maintain more regular blood glucose levels with fewer and lower spikes. Curiously, the algorithm even proposed counterintuitive suggestions, such as having ice cream at certain times (though only with moderation and qualification).[215] After the dietary intervention, the study participants observed healthier blood glucose levels prone to less spiking.

The researchers also used machine learning and data science approaches, feeding in dietary data and the outcomes. Over time, the program "learned" what foods elicited different responses based on food compositions, the gut microbiota, and attributes (e.g., body mass index) of study participants. The study's model was able to predict the glucose response better than the control model based on just the number of calories in the food. The Personalized Nutrition Project proved helpful; however, the insights remain individualized to each study participant. Every study participant's microbiotic footprint, attributes, and current glycemic responses must be taught to the algorithm before dietary recommendations can be made. The algorithm assigns weights to the various attributes, and given the interdependence, these attributes are not always transferable to someone outside of the study.

So, it's mildly disappointing that we did not learn more global insight from the study to extend to every person. The lack of universal wisdom may speak to just how difficult doing nutritional science is. It's highly personal as the study's title suggests. The authors also discuss at some length how our different microbiota affect the body's responses (though such effects are not as significant as the food itself). But there are also other factors that are far harder to capture in metrics that fit their model. In particular, I see the digestibility (structure of food) as having a sizable impact, which in and of itself is difficult to incorporate quantitatively in a mathematical diet model.

To go back to our metaphor, when, then, does a river overflow? It could be after a torrential downpour or when some downstream blockage occurs as when a dam is erected. Putting it plainly, it's how quickly the river fills up versus how quickly it empties. Diet obviously affects the filling up the most. And the depletion will certainly be a function of our age, microbiota, weight, regulation, and exercise habits. Interestingly, there are ebbs in our metabolic flow within the twenty-four hours of the day, as exemplified by The Personalized Nutrition Project. For example, our latent clock regulates our metabolism in accordance with our sleep schedule. Our glucose level spikes when we wake up.[216] Ultimately, I suspect this elevated blood sugar is to bring about wakefulness.

During our starvation-ridden evolutionary development, the biggest obstacle was liberating the components of consumed food so they could be metabolized. Compared to other organisms, we're more on the generalist end when it comes to diet. Bacteria, for example, can only consume specific compounds (e.g., sugar molecules). We can consume much of the same compounds as well as extracting them from more complexely-structured foods, such as fruit and cooked meat. To cope with the variation in sources, we evolved bags of acid (our stomachs) whose sole function is to wring as many calories as possible out of what we eat. The bags of acid, our stomachs, contain enzymes such as pepsin, designed to chip away at protein, as well as lipases to do the same with lipids. Metabolically, the limiting factor is how quickly we can turn food into suitable components. The digestibility, i.e., structure, of the food clearly plays into our diets.

For example, fruit and fruit juices have nearly identical caloric compositions. More than ninety percent of their calories are represented by sugar molecules. But in terms of how we metabolize them, the sugars within the juice are immediately accessible because they require no digestion before its metabolites enter the bloodstream. As a result, the river fills faster upon consuming the juice versus the fruit. There are certainly occasions where the juice serves us better than the fruit (e.g., to moderate hypoglycemia), but the whole piece of fruit will, overall, better help us manage the metabolic river because it is metabolized more slowly.

Foods with intricate structures, common to our ancestors' diets, such as whole vegetables and fruits, take our body some work to digest. The carbohydrates and nutrients can be freed, but our body has to work through the pectin and starchy structures. As noted, sometimes our individualized gut microbiota help us chomp through these hardy substances. Sometimes we cannot recover all of the nutrients. Ever consume whole corn kernels and check the toilet the next day? Today, with advanced food technology, we can sidestep the hurdle. We can imbibe sugary fruit juice and carbohydrate-rich beer that deluge our metabolic river, driving fat synthesis with ease.

We need to understand how foods, based on their structure, fill up our metabolic river. One limitation of the Personalized Nutrition Study was the reliance on measuring blood glucose. This glycemic index assigns a value to food based on how much it spikes our blood sugar. Ultimately, the glucose spike leads to a release of insulin, a hormone that signals the body to absorb blood sugar into liver and fat cells, overfilling those metabolic networks and promoting the synthesis of reserve fat. Glucose is also not the only way to trigger the release of insulin, for digesting protein-rich foods spreads arginine and leucine, which also stimulate the release of insulin. As a result, the insulin index will better represent the rising and falling tide of metabolism. So even meat, which is mostly fat and protein, will register insulin activity even though containing little to no sugar. As a result, the study might have overvalued meat for not spiking blood sugar as much.

If we consumed amino acids directly instead of protein, we would absorb and metabolize nearly everything.[217] On the other hand, intact protein takes at least a few hours for us to digest completely.[218] Furthermore, we know that different proteins can be digested at different speeds.[219] We do not have exact mathematical equations that can tell us how quickly we digest foods[220] because food digestion is ultimately the result of numerous factors such as the pH of the stomach; the structures of the food protein and the starches of the fruit/vegetables;[221] the viscosity of the food slurry that's being churned within our stomach; and the gut microbiota that help us digest food further. Food digestion is so dependent on a multitude of interconnected factors that it also may lie within a region of unpredictability, like weather (see Appendix A). We may never have equations that tell us precisely how we will digest a meal.

There is good news though. We do know that most of the variance in trying to develop our mathematical models comes from varying structures of the proteins and the starches. This brings us to the punchline: protein is a slow burn substrate. In other words, our stomach acids have to chop protein up into bite-sized chunks. We digest them slowly, enabling our metabolism to regulate with better health outcomes. It's the *structural* quality of protein that partly renders it valuable for our health. It's a battery that discharges slowly over time, keeping our insulin in check, when consumed in place of high-glycemic foods.

This slow-burn property isn't just limited to proteins though; fruits and vegetables also provide a similar challenge. We excavate nutrients from produce slowly, as the starches have to be broken down into their constituent sugars with the help of enzymes and our gut microbiota. Given the picayune protein that we actually need to maintain a healthy lifestyle, *fruits and vegetables are often suitable nutritional surrogates for protein-rich foods.*

Additionally, there is nothing special about animal protein versus non-animal protein for our health. If anything, limiting ourselves to only animal protein will prove disadvantageous in the long run. We can imagine a future where the fermentation-based protein is tailored and customizable. We could have different protein structures with varying digestibility.[222] For example, older adults with a diminished ability to digest could consume structurally flimsier proteins to counter muscle atrophy.[223] Younger people could consume hardier proteins that manage their insulin response better.

Customized proteins sound like something that would be "processed." Unfortunately, the word "processed" has become negatively associated with food versus "natural," which as we've seen has an equally fluid and opaque meaning. Like "natural" and "organic," "processed" on its own is too imprecise a descriptor, but we can improve it with a strong definition. Getting to the subtext, I believe the most charitable definition can mean food that becomes more digestible and accessible to metabolism compared to its original form. Therefore, chocolate candy bars are more "processed" compared to the original cocoa. Kale smoothies are more processed compared to the original leafy green. Beef and Beyond Meat burgers are more processed compared to the respective original cow flesh and peas. Sometimes, the processing is a bad quality. Often, eating oranges directly would serve better than drinking higher glycemic orange juice. But sometimes, processing serves us well, as in the prior example of older adults needing easier access to nutrients. We can also think of babies as needing more processed foods. Their digestive tracts don't accommodate whole broccoli florets but welcome them puréed. Finally, one of our oldest culinary inventions is entirely processed. Wheat cannot be directly consumed; the grain is just too hardy for our digestive tract. But when we process or mill it, we obtain flour that can be baked into bread. Bread can lose some nutrients during the milling, and sometimes these vitamins (e.g., niacin) are added back in as supplements.[224] This is not a bad thing.

SUPPLEMENTATION AND ADVERSE QUALITIES

As briefly mentioned earlier, not all of the essential elements our bodies need and derive from metabolism are fungible. We must obtain essential vitamins, such as B_{12}, and minerals from our diet. Our metabolism cannot produce these compounds from our enzymatic network. An obvious solution to this problem is to fortify or supplement our foods with these molecules. In the early 20th century, we noticed thyroid issues, specifically with ailments such as goiter.[225] The diet of the day didn't include enough foods containing iodine, whose absence exacerbated these public health issues. After a successful intervention study and pilot in Switzerland, iodized salt hit the United States grocery shelves in 1924. As a result, iodine levels have remained sufficient for over eighty-five percent of the United States population.[226]

Unfortunately, public perception of food fortification has in the modern era turned mostly negative. Consumers see fortification as rendering a food more "processed," packaged with negative connotation.[227] Also, many see fortification as a marketing tactic, not actually improving the salubrity of the food. These are disappointing, baseless views. We should welcome any opportunity to get the nutrients that we require any way we can and not mandate that they come from a "natural" source.

So far, there is no data to suggest that obtaining such nutrients from other sources has any deleterious effect on our healthy functioning. And why should we expect otherwise? Ultimately, the best explanation is that these molecules are chemically indistinguishable and independent of provenance. Certainly, different individuals may have differing abilities to absorb various vitamins. For example, consider vitamin B_{12}. Vitamin B_{12} is water soluble, meaning that when it is exposed it is susceptible to stomach acids. Within meat and animal products, B_{12} tends to be protected by association with proteins.[228] The proteins are cleaved off when the complex reaches our stomach, and then the vitamin associates to a protective compound to be eventually absorbed into the body.

Vitamin B_{12} famously does not appear in many foods outside of animal products. However, microbes naturally produce vitamin B_{12}.[229] And given the impressive metrics of fermentation processes, we can use bioreactors to produce vitamin B_{12} cheaply and quickly. I can purchase a year's worth of supplements for under $20. When I take B_{12}, I generally take more than the daily recommendation to account for the amount that cannot be absorbed. This seems to work out; anyone can obtain enough B_{12} with oral supplements.[230] Eventually, I would prefer B_{12} to be supplemented directly in the foods that I eat. I would even appreciate being able to consume the vitamin B_{12}-protein complexes that mimic those found in animal products. Nonetheless, I do not seem to be B_{12} deficient, nor do others on an animal-free diet who are conscientious about adequate supplementation.[231]

On the other side of non-fungible entities are toxins. While meat-eating advocates are quick to highlight the zinc, vitamin B_{12}, and omega fatty acids in animal products, what about the adverse elements within? Any holistic assessment must consider both the bad and good. For example, mercury poisoning primarily comes from eating fish.[232] Mechanistically, mercury inhibits vitamin-producing enzymes and the body's ability to generate antioxidants. Antioxidants promote brain function, which consumes proportionally high levels of oxygen. The oxygenation activity spontaneously generates reactive oxidative molecules that can wreak chemical damage. Antioxidants directly quench these destructive species. Altogether, excessive mercury consumption can hasten a person toward mad hatter's disease (also known as erethism, or mercury poisoning), named for the ill effects that followed the historical use of mercury in hat making. This disease is characterized by headaches, delirium, and hallucinations. Technologically, it would be difficult to detoxify fish of mercury without completely deconstructing the carcass. Mercury soaks through the viscera and muscles. With original foods and alternative protein, we have the ability to construct the flesh we deem fit. We would not need to cultivate protein from the sea that's been polluted with too much mercury.

Additionally, we have some associative, epidemiological evidence to link red meat consumption and colorectal cancer, but this has been challenged and disputed.[233] There are also murmurs in the epidemiological scientific community of increased cardiovascular disease and increased mortality related to meat consumption.[234] And likely, we will not be able to resolve the contention without mechanistic details for exactly how meat could create such issues (i.e., finding, testing, and incriminating the complicit molecules).

Uncertainty over the healthiness of animal products is part of the rub though, and a point in favor of a future without animal products. As discussed in the last chapter, all scientific characterization methods have limitations: they miss some things and overestimate others. It'll be laborious and unending to probe every element in a steak and assess how each affects human health. We also have no sense of the variance between different animal cultivations; so a specific species of cow may have flesh worse or better for our health. In contrast, our 3D steak-printer of the future will fashion food with constituents we are able to know down to every molecular detail. We would know exactly what proteins, fats, carbohydrates, and molecules go into each food, could map their structure, and have a better chance to associate them to human health outcomes. Anti-GMO lobbying forces often spout, "Shouldn't you know what's in your food?" as an argument for mandatory labeling of GMO foods. Ironically, the argument also favors the continued use and development of genetically modified foods themselves. We know more about a GMO food than a non-GMO one because we were intentional about the design toward a desired outcome.

So far, I haven't actually addressed the visceral wariness of eating GMOs. I acknowledge that eating something foreign and of industrial origin brings to mind a dystopian image of some mad doctor injecting a glowing green liquid into our bodies. This is why Impossible Foods and Beyond Meat spend millions to make their products look like the animal-based analog and why they're displayed next to animal meat in grocery stores. But in this chapter, we gained some solid insight into the details of nutrition; in particular, we know that food is ultimately broken down into molecules, which are then metabolized. Now, we can shoo away the albatross because we know more and can color GMOs with more nuance.

THE NUTRITION OF GMOS AND ORIGINAL FOODS

I remember the brief heyday of olestra from my childhood. This was a tripartite fat, a triglyceride, whose chemical structure resembles a trident. In olestra, the backbone structure was changed from glycerol to the sugar sucrose. It cooked and tasted like oil, and even better, it wasn't metabolized, so it contained zero calories. Olestra was approved in 1996 by the Food and Drug Administration as a food additive, and in 1998, Proctor and Gamble (P&G) started introducing a number of products with olestra and marketing them heavily.[235] I still remember walking through grocery stores with my dad and seeing "Olestra" plastered on bags of chips everywhere.

But there was a catch. Olestra was known for inducing abdominal cramps and for loosening stools, in a charmingly known side effect referred to as "anal leakage,"[236] (though these observations have been challenged with a formal randomization study).[237] In a separate study, olestra was suggested to be detrimental and linked to weight gain.[238] Rats that ate both food with normal fats and olestra gained *more* weight than rats eating just normal fats. The authors hypothesized that upon eating olestra, the body's nutritional regulation went haywire. The brain received misleading signals from the metabolic cycle, implicitly perceiving the consumption of fat, but, upon breakdown, the fat was missing. This got the biological alarm system blaring, telling the brain it was not satiated. The mechanistic details here are still being investigated. One lab-based hypothesis suggests that the ringing alarm induces mice to eat more, and another that the alarm redirects the metabolic rivers toward fat storage. The same effect has been implicated in the consumption of artificial sweeteners.[239]

The olestra case study may inspire further reluctance to consume GMOs and orthogonal foods, so I wanted to clarify this. Olestra is *chemically* foreign, meaning that our biology never evolved to handle it. The molecular structure is not found in nature. Chemical orthogonality even scares an avowed anti-naturalist such as me. When such foreign chemicals are consumed in abundance, as they would be in food, we have no idea about how the unfamiliar chemicals will interact and the effects they'll have within our body. In fact, understanding the interactions of molecules and the rest of our body remains a nontrivial, arduous endeavor in biology research.[240]

However, nearly all GMOs and potential alternative foods are chemically native and familiar in that, once broken down by the body, they're indistinguishable from non-GMO foods. Such foods are really just a new configuration of the same molecules that we've already been eating throughout our evolution: amino acids, sugars, fats, minerals, and vitamins. One way to counteract GMO neophobia is to admit that there's no need for an olestra-like, foreign chemical creation.

The quest for alternatives like olestra and artificial sweeteners began with food scientists who sought to provide quality eating and drinking experiences to people without the nutritional detriment of actually consuming fats and sugars. I laud the intention behind this pursuit. However, I'm not convinced that just slaking hedonistic pleasures is the most durable, productive strategy when it comes to creating future food. We would be remiss not to consider teaching people how to manage their hedonism and mutate their valences, i.e., what sensory inputs they associate positively and negatively. So, let's go to the next chapter and see how that plays into a future after meat.

CHAPTER TERMS

- **epidemiological:** an approach to scientific research that draws inferences from uncontrolled events
- **digestion**: a mechanical and biochemical process to break down food into metabolizable constituents
- **metabolism**: a biological process to chemically convert molecules into mass and energy
- **regulation (biology)**: the act of biological systems responding to internal or external changes
- **fungible (metabolism)**: the exchangeability of large components of metabolism: Protein can substitute for many dietary functions usually served by carbohydrates and vice versa.
- **obligatory nitrogen loss**: the minimum amount of nitrogen (i.e., protein) needed per day
- **conserved**: the repetition of biological genes from one species to another

CHAPTER SUMMARY

Nutritional studies are difficult to perform and not always well-conceived by researchers. Sometimes controls are difficult, effect sizes are so small as to be almost immeasurable, or the posed questions are unreasonable or practically unanswerable. Nonetheless, we still know a great deal about nutrition from groundwork in metabolism. The food we consume is digested into molecular entities, and these entities enter metabolism, a network of chemical transformations. Given their regulation and interconnectivity, entities within metabolism can be viewed to have a high degree of fungibility, meaning that the root nutritional sources of much of what we're eating—ground beef or an Impossible burger patty—are interchangeable. The value of protein is questionable if it can be substituted with fat or sugar. And we find that the amount of protein we need falls well short of current recommendations when we use a rigorous mass-balance approach to calculate it. Instead, the value of protein seems to be related to its slow digestion and releasing of nutrients steadily into our metabolism. The same effect can be reproduced with vegetables and fruits. Additionally, we should not be afraid of vitamin supplements seeded in various foods because they're chemically identical to what we gain from food itself. As we evaluate foods, we also must consider what's adverse in them. We can't extol the health benefits of animal-based foods without considering the downsides (e.g., mercury poisoning) or confessing our utter lack of knowledge of the details. Finally, the fencing off of genetically modified organism (GMO) food technology is misdirected. GMO food generally has the same ingredients we normally eat, though they may be structurally arranged in a different manner.

CHAPTER SEVEN

Hedonism and Food

MORE REFINED TASTES

One of my favorite videos on the Internet is an advertisement by the Finnish company Fazer.[241] Fazer is the country's leading producer of *salmiakki*, a candied, salted licorice. They formulated a *salmiakki*-flavored ice cream, and the video is promoting this new product. For the uninitiated, *salmiakki* will likely taste as objectionable as it sounds: it's bitter, cacophonous, and foreign, but the Finns love it. The first half of the video features interviews with non-Finns, who are tasting *salmiakki* for the first time, only to quickly retch in disgust. When probed by the video producers, they immediately inveigh, "Awful," "This is the worst ice cream I ever had," and "I just can't." Naturally, the latter half of the video trots out Finns eating the ice cream and exclaiming: "I don't know how you could make it any better; it's already so good," "If I had this in my fridge, it would not stay there for long," and "Can I have another one?"

All the Nordic countries—Sweden, Denmark, Iceland, Norway, and Finland—love salted licorice. Go into any gas station in those countries and you'll find salted licorice products adorning the walls, shelves, and counters. How is it that a product can be so beloved in one area and reviled outside it? We might instinctually believe that this is due to ancestry differences. However, the video highlights a Finn of African descent among the aficionados. Furthermore, my friend of Indian descent loves salted licorice, after having lived in Denmark for some time. And I too, who am also of Indian descent, have slowly come to appreciate *salmiakki* over time after some prodding and trying more varieties.

This dichotomy of *salmiakki* lovers and haters demonstrates that what we consider to be a "good" sensory pleasure can mutate over time, and it's hardly the only example. Most of us remember getting used to beer, likely finding it bitter and offensive on first sip. But we can form a liking quickly, especially as social norms inculcate that preference into us. Bitter foods, in particular, seem to follow this script of initial determent to eventual delight (think coffee and scotch). In the obverse, what we consider good may eventually turn gross. I gleefully consumed beef burgers as a meat-eating teenager but gave up meat when I got to college. While living in Switzerland, I heard the news about the Impossible Burger and just how well it captured the taste of beef. When I moved back to the States, I had my first opportunity to try one in the summer of 2018, a full ten years after forsaking meat. As I chomped into the burger, I found it off-putting. The heme-iron taste reminded me of biting my tongue; I didn't find it pleasant even though that's how beef burgers apparently taste.

Most of us intuit that our tastes can change; otherwise, we would still have the palette of kids and prefer eating macaroni and cheese and chocolate pudding. This feeling of positivity and negativity associated with unique experiences has a name: **valence**. Positive valence is often associated with a flash of dopamine throughout our brain, reinforcing our preferences further. There's research that invokes dopamine to explain our predilection for animal products: one theory asserts that our ancestors, *Homo habilis*, developed increased dopamine levels after meat consumption 2 million years ago.[242] Ice cream, rich in both fat and sugar, can trigger a dopamine response, even surpassing that of cocaine.[243] The dopamine-coupled pleasure is the third of the four Ns of eating meat: nice. The positive valence is hard to give up, and this stymies the transition to a future without animal products. We imagine a dreary, subdued life without flavor. However, as we, albeit superficially, explored, valences can mutate, and our **hedonism** can transform.

We've seen some of this in the transition from animal to non-animal products. Dairy milk has seen a substantial decline in consumption over the past twenty years in the US market. The average US citizen consumed nearly 100 liters (26 gallons) of milk per year back in 2000, but by 2019, that figure had sunk to 70 liters (18 gallons).[244] In its stead, almond, oat, and soymilk have surged, accounting for fourteen percent of the market, with most of the growth being very recent.[245] This growth is not driven by increased availability. In fact, the sales of plant-based milks are exceeding the pace of their distribution, meaning customer preferences are driving the curve. Indeed, some studies echo favorable taste evaluation on the non-dairy milks, particularly the sweetened ones.[246] Now at my local Starbucks, I have at least three varieties of non-dairy milk—soy, almond, and oat milk—from which to choose.

While I'm confident that, in the long run, plant-based foods will taste vastly better than their animal counterparts, there's another consideration. Directly addressing and modifying our hedonic programming *should* be not only considered, but taught and fostered broadly in the population. When life goes poorly or we suffer or feel anxious, we should not always expect some external pleasure to rescue us; be it a box of ice cream, a different partner, some new product, a vacation, etc. Sometimes, we're better served to internally reprogram and develop our resilience to handle vicissitudes and seek new habits and training. Examples could be joining WW™ (Weight Watchers), or engaging in therapy, yoga, or meditation practices. With better understanding of our brain circuitry, we should be able gain insight into the hedonistic pleasures we turn to that don't serve us or the planet and make different choices.

HEDONISM IN OUR LIVES

Being able to note—and remember—a sensation as good or bad has obvious evolutionary advantages. Our brain has evolved to repudiate offending smells and bitter tastes as their noxiousness alert us to toxins in spoiled or dangerous foods. The bitterness of natural almonds clues us in to their highly poisonous cyanide content.[247] Our ancestors learned through trial and error that the sweeter variant, or **cultivar**, of almonds was safe to eat, and this is what we now breed exclusively throughout the world today.[248] High caloric foods—sweets and meats—enliven our taste buds and light up the pleasurable neural pathways in our brains, which, over generations evolved to give us a preference for them, while at the same time programming an aversion to certain dangerous or possibly toxic foods. In this way, our taste buds helped to ensure the continued survival of the human species.

The way we characterize sensations depends on our neural circuitry. Neurons extend out from a brain region, called the tegmentum, to another brain region, the nucleus accumbens.[249] When we try a new, delicious food for the first time, pleasure senses trigger the start of the Mesolimbic pathway, where the neuron activation cascades, ultimately resulting in a spike of dopamine levels. The dopamine binds to receptor proteins of learning-associated brain cells. We crave that experience of pleasure again, and our minds link the events to the precipitating causes. The net result is that we are conditioned to continually pursue such behavior, in this case, seeking out more of the pleasure-inducing food.[250] After we've tried the food for the first time, we're conditioned to want it, specifically, again. And the next time around, the dopamine spike occurs in *anticipation* of consuming the food. This dopamine cascade applies to other body senses, as well, even without the prior experience; for example, sexually naïve males will release dopamine even at the prospect of sex.[251]

On the other side, dopamine molds our psychology to avoid painful experiences, and to actively develop aversion. Dopamine-releasing, or dopaminergic, neurons normally leak out a steady stream of dopamine, even when stimuli are absent. When we experience a sharp pain on picking a thorny flower, this steady stream of dopamine slows or stops abruptly, and our brain circuitry computes this gap in dopamine level as negative.[252] As a result, we learn to avoid the complicit situation that led to the response.

In sum, dopamine can be understood as a mechanism for **motivational salience**, that is, modifying our behavior toward biological, evolution-developed goals. At the end of the day, the effect of evolutionary programming had nothing to do with how we felt, it simply motivated us to accomplish certain tasks. Any positive feelings or associations became an instrument toward those ends, and continues as such.

We do not even have to be mindful to find motivational salience in action. How many times might you wake up to the moment and find you have unconsciously been scrolling through Facebook, Instagram, or Reddit? I've even found myself mindlessly trying to navigate to Reddit while being on Reddit. When this dopamine system goes deeply awry, we're prone to addiction, whether it be drugs, alcohol, or something else that aberrantly reinforces dopamine-release patterns. Social media companies play to the dopamine response to ensure that we stay transfixed to their platforms. These companies reap advertisement revenue commensurate to the time that we park on their platforms. Getting Likes, internet karma, or even just looking at pictures of cats can stoke our streams of dopamine, giving rise to a pleasure response that we want to repeat, which ultimately fills these companies' coffers. Eyeballs equal advertising money. YouTube's recommendation algorithm is a perfect example: enticing videos are predicted, selected, and surfaced to the user, and the platform will automatically play to the next one so that the viewer remains, sometimes unaware, and thereby drives up YouTube's advertising revenue.[253] Indeed, 70 percent of YouTube's views come through the recommendation engine.[254]

So, if dopamine release itself does not directly equate to happiness or pleasure, are we truly experiencing pleasure by surfing Facebook? Likewise, we can be unconsciously driven toward food or away from intimidating social situations. We have feelings of pleasure that can result in the release of dopamine. But ultimately, dopamine is the *result* of the actual pleasure sensation. In real terms, we could conceivably modify the behavior or motivational salience of someone by administering and depleting dopamine; however, just administering dopamine itself would not generate happiness, as was confirmed by experiments in the 1980s.[255] The lack of pleasure is not cured by increasing a person's dopamine levels.

Nevertheless, *in certain situations*, we can correlate a person's or animal's pleasure to the release of dopamine, given that dopamine is often released concomitant to the sensation of pleasure. And the magnitude of dopamine should correspond to the magnitude of pleasure sensation, too. There are limits to this because it's truly difficult to measure pleasure; the problem even has a name: "The Hard Problem of Consciousness." We currently have no way of "measuring" someone's conscious state in an objective, universally applicable manner, e.g., using a mass spectrometer or some chemical assay. We can't tell if someone is experiencing pleasure nor to what magnitude in comparison with someone else. We don't know if they're lying, mistaken, or how their experience compares to someone who verbally professes feeling pleasure with the same experience but may not have the same feelings. But tying dopamine release to pleasure is one of the best methods we have to understand human hedonism, and per the Popperian paradigm posed in Chapter 2, that means we must accede to it unless and until we have something better.

Neuroscientist Robert Sapolsky in his book, *Behave,* highlights some disheartening findings about dopamine release and activity.[256] A monkey learns that he will receive a delectable raisin after pressing a lever ten times. On the first success, the monkey experiences a wave of dopamine, stemming from the initial pleasure and surprise activation.[257] The monkey assigns a positive valence to the raisin. Then on a second trial, the monkey receives two raisins, resulting in roughly double the dopamine release, or for our purposes, double pleasure. In the consequent trials, the monkey continually presses the levers and receives two raisins. However, the pleasure response diminishes as the monkey habituates to the pleasure and the lack of surprise. Disconcertingly though, eventually the monkey can end up experiencing *less* pleasure than when he or she received that one raisin on the initial trial. The monkey can end up less happy receiving two raisins then when he first received one raisin.

When we understand pleasure and dopamine from an evolutionary perspective, this shouldn't surprise us. Complacency is a bad trait: we won't be able to be healthy enough to produce new members of human society by remaining in the same place should a volcano erupt and wipe out the food in our region. We improve the survival of the species by constantly probing for new opportunities and new pleasures, even if that sometimes makes us more miserable. Consequently, our pleasure is designed to abate and diminish for the same repeated action. Two raisins is fine, but what if this other lever grants us three raisins? Altogether, pleasure and dopamine are *relative* and answer the question, how much better is this *new* event versus the current expectation?

This effect, termed the **hedonic treadmill**, is well known. The consumption of the raisin eventually becomes perfunctory, and the resulting pleasure will continue to abate with each consequent trial. (Though, if the raisin is rancid, and the monkey retches in disgust, the dopamine wiring may switch polarity to the opposite direction.) Certainly, we can refactor pleasure by spacing it out over time, as professor Daniel Gilbert highlights in his book, *Stumbling on Happiness*. If we have our favorite food every day, then it will likely cease to be our favorite food. If we space our consumption to every two weeks or every month, however, then we can reap the same pleasure in eating it over time.[258] The concept of the hedonic treadmill bears out with other data. For example, once individuals reach a certain level of income, around $75,000 per year in the United States (as of 2010), their happiness levels start to flatten as their income increases.[259] Making more money does make us happier, but not in the same proportion. Our happiness would jump more if our income doubled from $75,000 per year to $150,000, than from $150,000 to $300,000.

Thankfully, the ability to habituate eases negative valence pursuits as well. In other words, in repeating behaviors we find irritating or noxious, like cleaning the house and exercising physically every day, we lessen that perceived displeasure. For example, having our children brush their teeth for the first time is like pulling teeth. But after much prodding, pleading, and pleasing, children can adopt a consistent teeth-brushing regimen without further intervention. Like clockwork, we learn to brush our teeth at night before we go to bed and again in the morning. We can cultivate similar approaches to jogging or washing the dishes. What initially was a repulsive activity eventually becomes nondescript or perhaps even enjoyable. Initially, we may have that dopamine drop, signifying displeasure. The drop, however, diminishes with regular activity. We can adapt or even eventually associate pleasure with the task.

So that brings us to the pleasure of consuming animal products. Clearly, our neurobiology values the expedient, caloric-richness of animal flesh, which has helped fuel our species' proliferation. A number of companies and researchers have persistently studied[260] the distinctive aroma, and several, among them Impossible Foods, have spent gobs of money trying to reproduce an authentic meaty aroma in their veggie burgers.[261] The company's thesis is that we should be able to reproduce the experience and pleasure of consuming meat in the consumption of non-animal products. In fact, many of the actors of the alt-food, alt-protein space believe that we'll be able to reproduce or even exceed these hedonic pleasures.[262] I agree, especially given the intractability of animal technology (Chapter 5), and I will argue such in the next chapter. Even more, I think we'll find *new* tastes and types of pleasures currently not available or possible with animal technology. And, yes, I am hinting that this might be a more fruitful path for these alt-food companies. Let's punt on that point until the next chapter though.

For now, let's discuss the other meta question here: is it continually sustainable to look for greater pleasure after greater pleasure? What about changing the way our minds work and our relationship to the foods that we seek for pleasure? We need to step outside of the paradigm and ask if there is a better way to enrich everyone's lives. Hedonism and pleasure are tied to how we experience and perceive the world, specifically how we consciously perceive it.

CONSCIOUSNESS

Why are humans **conscious**, that is, having an awareness of existence and a physical, outside world? Throughout the book so far, we've argued that any naturally occurring biology can be explained in evolutionary terms. So, we must have an evolutionary reason to have developed such a sophisticated, vivid system to grok the world. As far as we can tell, lower organisms such as bacteria do not share the same capability that humans and larger animals possess. Bacteria whip themselves toward sources of food based on simple, well-studied **chemotaxis** systems to arrive at the beloved sugar in their environments. They can detect differences in concentration of sugar around them, and of course, they gravitate toward the areas with more sugar.

Bacteria simply have to know where the food is. Computing more sophisticated functions is not worth the cost under the burden of the Pareto frontier. In contrast, humans are non-aquatic generalists. While smells can lift us toward desirable food, most biological goals are more complicated for us. Suppose that a hunter-gatherer ancestor sees ripe bananas hanging in a tree. Her legs and feet would need to propel her to the base of the tree. Once there, she must evaluate a path to climb up to the branches that bear fruit. Her arms and legs must then coordinate to traverse up the tree. Finally, she must find a safe, facile approach to get back to the ground again.

This entire activity of climbing up to grab the bananas is the coordination of many subsystems within the hunter-gatherer's body: an intellectual evaluation of the best path forward, motor cortices to direct the movement of arms, wrists, and elbows, among others. Even parts of the brain not directly under her perceptional control need to be privy to the situation, so the cortex responsible for breath modulation may automatically oxygenate the muscles so that they don't prematurely tire. Her breathing patterns may even change without her being aware, similar to how our breath quickens after exercising or doing something strenuous. We don't direct our body to breathe faster; it just happens.

The hunter-gatherer clearly needs a way to broadcast all of this information—the banana being up in the tree, the potential route to climb up, when to rest, etc., to the relevant parts of her body. She needs a central bulletin board accessible to all of her various subminds. She needs consciousness.[263] This staging area analogizes to a conference of board members that determines the long-term, sweeping actions of a company. It may sound absurd to say "minds" when referring to one person, but a multiple minds model accords with neuroscience research.[264] Our minds are layered and hierarchical, comprised of **homunculi**, or smaller minds. For example, the motor cortices of the brain can be somewhat physically demarcated via electrical stimulation experiments.[265] Regions of the brain can be associated with control of the tongue, elbows, etc., though there is significant overlap, and jurisdiction for control can transfer after injury.[266]

Beyond just our motor cortices, there is evidence that the thinking part of the brain also subdivides further, an idea suggested by the thought-provoking, split-brain experiments. As background, brains divide between the left and right hemispheres. If we close our eyes and visualize a brain, these are the two lima beans glued together. Each of these hemispheres has been associated with particular thinking functions. The right hemisphere controls the left hand, whereas the left manages the right hand. The sole connection between the hemispheres, a structure called the corpus callosum, was severed in patients as a potential treatment for crippling epilepsy in experiments led by Roger W. Sperry.[267] Our knee-jerk reaction is that the treatment would be disastrous, but the patients demonstrated normal behavior after surgery. Sperry and team could probe each hemisphere's thought processes individually, for example by asking the messenger hand to write out something (e.g., left hand for the right hemisphere). As philosopher Sam Harris writes in *Waking Up*, a young split-brain patient was asked what he wanted to be when he grew up.[268] His left hemisphere verbalized out loud, "a draftsman," but his right hemisphere brain spelled out "racing driver."[269] Harris points out:

> *In fact, the divided hemispheres sometimes seem to address each other directly, in the form of a verbalized, interhemispheric argument.*[270]

The two hemispheres seem to differ in temperament and seem to have distinct personalities, distinct consciousnesses. The corpus callosum bridges the two to create a unity and integrates both into the same boardroom, the same consciousness.

The corollary is that each hemisphere actually subdivides further. If there were ways to separate the equivalent corpus callosums within, we would obtain even more fractured subminds (**fractal**). And each submind would have its own consciousness.[271] We don't have a way to actually sever such connections nor would that be ethical or advisable, but this already plays out in a fully interconnected mind. When an input first presents itself, each submind has its own congress, each of those elevate to a higher congress up the chain. Some conjecture-refutation mechanism exists, determining whether a thought gets elevated up the chain. The thought may be modified before reaching the consciousness that we as individuals experience.[272]

This is why we can solve a problem long after we've consciously stopped thinking about it. Have you ever had a hard problem from work, school, or home life for which you couldn't immediately propose a solution? But then, you start mowing the law or you take a shower and a clever, enveloping solution just wallops you? This revelation is our subminds at work. Their actions aren't privy to our main conscious boardroom, but they're present and dutifully working. Even as I write this book, I notice that I hit a writer's block after a few hours of working. But if I come back a day or two later, even without actively working on or thinking about content, my creativity reservoir is full again, and I write out fluidly or easily solve a problem that utterly flummoxed me before.

And as discussed earlier, our conscious perceptions do not have to exactly accord to reality. They merely have to be evolutionarily useful. For example, we all have a blind spot in our eyes, but it is unbeknownst to us because of the way we process vision mentally: if we close one eye and look straight ahead at an apple, then move ourselves close enough to the apple, the apple will suddenly disappear. There are many websites demonstrating this phenomenon.[273] The Pareto frontier means that we're only going to perceive reality to the point of functional, evolutionary usefulness and no more; our consciousnesses are not going to be optimized to grapple abstract, mathematical concepts. As we saw earlier, the biggest disconnect occurs as we push the boundaries of physical science. In

Appendix A, we discuss the Multiverse theory as the best explanation for quantum theory physics. Biologically though, we are not able to perceive reality this way, and accordingly, the Multiverse remains a "fringe" theory.

Our main consciousness boardroom can only handle one thought or sensation at a time; this limitation prevents us from breaking the hedonism trap. As mentioned, sometimes this occurs without awareness as when I'll find myself grabbing a snack before even having a chance to stop myself. Given that our evolutionary biology needs to assess our actions, we experience a resulting feeling, such as "wow, that dark chocolate just enlivens my soul." We often caricature into automatons that carom from one rousing activity (e.g., video games) after another (e.g., fast food) to then crash (e.g., burnout from work) and to seek reprieve from the unhealthy feelings associated with the trap (e.g., vacation). Do we want life to just be a series of sensations and feelings, i.e., remain in the hedonism trap?

In the long run, augmenting and fixing human psychology deserves our attention. I am not merely arguing that we continue to do psychological and psychiatric research on human behavior. There's another important step: *we'll want to remediate "natural," evolutionary tendencies*. As stressed throughout the chapter, the way evolution shaped our psychology is purely instrumental toward the proliferation of our species, not adapted specifically to solve global problems, generate knowledge, nor make us happy, though those may incidentally occur. Many psychological traits are more problematic than helpful today.

Let's just assume for the future's sake that we don't want to be solely guided by feelings and reptilian automaticity; instead, we want to be guided by considered intention. Consider the power of acting in complete intention: this superpower would immediately kill every diet fad. We would all eat better. And regarding animal products, this means that the "nice" part of the four Ns would no longer hold so much sway over us. People wouldn't be steering into a McDonald's drive-thru on a dopamine whim. It'll be easier to give up animal products or to switch to replacements. It'll be easier to herald a healthier future after meat.

Even though I think humanity will learn how to surpass its outdated evolutionary psychological baggage, I expect change will be met with resistant headwinds from nearly all consumer-related companies. As we saw, Facebook's and YouTube's revenue is fueled by some of the worst evolutionary traits of human psychology. Fast food companies use many tricks in the dopamine playbook: addictive sugar, providing goods at speed and convenience, and using branding to associate pleasure in your brain.[274] And every consumer-facing company benefits from selling more, not less. Therefore, I'm not expecting a startup or company to "innovate" ourselves out of the hedonic traps, as that's a solution seemingly against every company's directive and bottom line. The solution to this problem must come from outside the private sector.

Thankfully, we can lessen these impedimenta and cultivate acting with intention using the practice of meditation. Meditation fosters intentionality by developing one's mindfulness, the ability to precisely interrogate one's consciousness boardroom.

MEDITATION

Unconscious behavior is diminished by becoming less mindless—doing things out of habit or on autopilot—and is also known as being more mindful. If I can gain more perception of reality as it is in the moment (being more mindful), the break from the constant discursive narrative in my head grants me the space and time to intentionally decide my next steps. If we take snacking for example, if I can perceptibly feel the cue from my body that I'm craving cheese on crackers but I'm not really hungry, then it's easier for me to just say no. If I notice my anger rising due to the actions of a coworker, I can steer away from saying something nasty and instead consciously decide to pursue a constructive outcome. Thankfully, gaining the ability to be mindful—to be more aware of our present state and surroundings—is akin to strength training. It's like a muscle that can be exercised and nourished to increase functionality and daily ubiquity.

We can cultivate more **mindfulness** through the practice of meditation. During a meditation session, one sits still and tries to focus on the breath, a mantra, or some other bodily sensation. Often the focus on the inhalation and exhalation is easiest for new meditators. Greenhorn meditators will notice that this is not easy: the mind will demur and derail, thinking about deadlines at work, a conversation with a loved one, or an errand to run. But as the meditative "muscle" strengthens through repeated exercise, then an adept meditator will develop higher levels of **metacognitive awareness**, and she will be able to maintain minutes of intentional presence in the moment. With even more practice, the minutes become hours. The meditative presence additionally seeps into the rest of her life. For example, her stomach might grumble, and her mind will release dopamine at the thought of grabbing a salami sandwich, but her meditation practice will remind her this is just a thought, that it will come and go, and she won't.

Metacognitive awareness is direct perception of the contents of consciousness, or being able to see the boardroom wholly and attentively. Observing the cascading of the breath, a sensation of warmth that runs over our back, or anything else that the different subminds conjure up to the primary consciousness. Even the arising of thoughts and their supplanting can be observed with enough training and experience. With even more training, sensations may bubble up without descriptor, e.g., a sharpness in the legs. Instead of reacting with a pain reflex, an unfiltered white noise might make its way to the consciousness boardroom for presentation without any narrative bundled to the feelings. Therefore there is no dopamine baggage to initiate a habituated response.

On the path to increased metacognitive awareness, meditators must develop **equanimity**, or the ability to attribute neutral valence to sensations. A painful sensation can be recast as just sensation. An urging to look at email can be acknowledged, and the associated tingling or tightness in the chest can be observed. A visually striking example of equanimity comes from the image of the Vietnamese monk, Thích Quảng Đức, who self-immolated in protest of the Ngo Dinh Diem regime (**Figure 16**). Journalist Malcom Browne captures Quảng Đức sitting upright, cross-legged, while maintaining perfect equanimity; even though Quảng Đức experiences what's often described as the most painful way to die,[275] Quảng Đức is still and steadfast in protest thanks to his Olympian-level meditation training.

Figure 16. The power of advanced meditation. Thích Quảng Dức self-immolating without apparent suffering in protest of the Diem regime on July 11, 1963.[276]

Equanimity is not the same thing as numbness. It's not the elimination of feelings; meditators still feel happiness and pain. Instead, it's being able to reframe sensations as nothing more than existing and present. While my meditative abilities have a long way to go, I've been able to consume raw, bitter spinach leaves more easily by just focusing on the sensory inputs and summoning awareness and equanimity. In the same way, I can quell deleterious desires: I don't have to eat cheese or spend money on frivolous trinkets on a trip just because my reflexive mind thinks I should. My mindfulness is omnipresent enough that I find social media mostly banal because I have less automatic attention-surrender. But I'm not emotionally bereft; I can still feel happy and motivated. At the same time, I'm also susceptible to procrastination and trepidation in writing this book, but those negative attributes have palpably alleviated and diminished in my life with ongoing meditation practice.

The net effect of meditation is being less enslaved to thoughts, ephemeral feelings, and valences—all of which underlie suffering, as posited by early Buddhists. They observed and made explicit, falsifiable claims about the mind and the inner workings; for example the subminds and the various congresses of consciousness.[277] These doctrines have been iterated over the years, thankfully with variants denuded of the unfalsifiable, supernatural overtures. And today, many resources exist to help one cultivate a meditative practice, including a litany of books, guided meditation, retreats, and mobile applications.

In the last section, I argued that having the ability to do exactly what we intend would be one of the most useful abilities realistically imaginable. So far though, there is no pill we can take to magically produce action, though we can get closer with meditation practice. Therefore, I claim that if one wants the increased ability to act according to their intentions, they should cultivate mindfulness.

And right now, the best way to develop mindfulness is through meditation practice. I claim that *meditation is one of the best activities that anyone can start right now.*

Yuval Noah Harari, furthermore, talks about the elements of novel experiences,[278] for example, why we seek out adventures. Harari proposes two variables: first, the novelty or stimulation of the experience itself. Second, our ability to fully take up the experience; for, if lost in thought or distracted by feelings, then that's less experience imparted to us. More mindfulness enables us to imbibe encounters more fully.

This means that we can reap more satisfaction from our trips or spending time with loved ones. Also, it means that we suffer less. The construction noise outside of your apartment no longer seems grating; it just fades into the background. And germane to the book topic: fleeting, hedonic pleasures such as those derived from certain foods will possess us less. Inherent hedonism will continually lose steam as humanity cultivates its mindfulness abilities. Given the primary benefit of mindfulness, i.e., having more intentionality in our lives, I argue that *we (humanity) will want everyone to practice it.* We'll see commitment to meditation akin to how we view devotion to physical exercise today: as an activity largely and categorically beneficial to anyone's wellbeing and something your doctor will routinely prescribe.

The only drawback with meditation that I see is the time required to cultivate it. Certainly, this time could be spent doing other productive, palliative pursuits such as physically exercising, learning a new skill, or volunteering, but I suspect for the most part meditation might replace or supplant activities that we approach for respite, such as watching television and engaging on social media platforms. In my personal experience, meditation practice has abrogated my desire to play video games. I cannot summon that experiential fusion that I had with video games as recently as three years ago, where I would lose myself in the world of the game and was ensorcelled for hours. I view this as a good thing—ten to fifteen hours of video game time per week has been supplanted with five to seven hours of meditation. The dividends of that investment have yielded more time and mental wherewithal to attempt more fulfilling projects—such as writing this book.

Asking everyone to meditate is easier said than done, and I'm in a privileged position to cultivate my own practice. My recommendation might be like that out-of-touch rich guy who implores everyone to take an extended vacation in the Maldives every year. However, I argue that we want *everyone* in the world to have more mindfulness, similar to how we want everyone to be healthier and to suffer less. Perhaps a nootropic drug can eventually conjure the same effect as 10 thousand hours of meditation. That would be a worthy endeavor, and I hope that we eventually figure out a facile, scalable way to improve everyone's mindfulness. But until that point, I will beat the drum that some meditation activity is better than none, and no one is remiss for having *more* mindfulness, all things being equal.

In terms of hedonism, I've harped on the advantages of meditation and mindfulness toward reducing humanity's reliance on animal technology. There's another mechanism at play though toward the underlying topic of this book: the generation of knowledge. In Chapter 2, we discussed how humans are natural knowledge generators. Our thinking process mirrors that of knowledge generation by conjecturing and refuting ideas.

When we have more mindfulness, unhelpful thoughts—grievances, possessiveness, egoism, and jealousy—capitalize and overcrowd our consciousness boardroom less frequently. This allows *new* thoughts to crop up for consideration because, remember, our boardroom can handle only so much at one time. Some studies highlight the ability to observe and attend (as granted by meditation) to perform better on creativity-oriented tasks; for example, children who performed mindfulness-boosting activities created higher-rated art versus the students in the control group who had no mindfulness activity.[279] In general, mindfulness has been associated with increased concentration,[280] reduced self-consciousness,[281] and increased openness to new experiences.[282]

The consequence, of course, is we'll be able to generate knowledge faster and more easily, and better ideas will displace lesser ideas more quickly. As claimed throughout this book, I see knowledge generation as the largest determinant toward a future after meat. If we globally improve our innovation capability, we'll reach that point faster.

Getting back to the hedonism argument: you may still be skeptical whether humanity can learn to control or wield hedonism more effectively, and for that reason, you think animal products are here to stay. Earlier, I hinted at the fact that non-animal products will be more hedonically pleasurable in the long run. In the next chapter, we'll explore this argument fully and completely dispense with the "nice" in the four Ns.

CHAPTER TERMS

- **valence:** the association of positivity or negativity to a sensory input
- **hedonism:** the pursuit of sensory pleasure
- **motivational salience:** programming of an individual toward an objective or perceived outcome
- **hedonic treadmill:** continual diminishment of pleasures as they occur more frequently or become overshadowed by more potent pleasures
- **consciousness:** a central bulletin board/conference room/ staging area for many subminds to access relevant information about the environment or bodily state
- **chemotaxis:** possibly construed as very proto-consciousness in microbes. Specifically, it's the detection of changing chemical concentrations to help microbes locate food.
- **homunculi:** historically referring to a small human. Here, it's used to mean subminds, each lacking the full functionality of a complete human mind.
- **fractal:** self-similarity that occurs at different levels. For example, zooming in on a snowflake's edge reveals a patterning similar to the body of the snowflake. Evidence such as the Split-Brain experiment suggests the possibility of our brains being fractal.
- **mindfulness**: awareness and cognition of the current moment
- **metacognitive awareness**: mindfulness of current thought processes and entry into the consciousness
- **equanimity:** being able to attribute neutral valence to a sensory input, thought, or anything else relayed into consciousness

CHAPTER SUMMARY

What tastes good to an individual is a subjective experience associated with a positive valence. That subjectivity can vary over time or in context: what was good can sour and vice versa. We can understand positive and negative feelings, including those coming from food, as evolutionary programming to modify our behavior—motivational salience—to meet the evolutionary proscribed objectives including pursuing sex or eating rich foods. Every behavior and action is evaluated relatively. Is a new activity better than the current one? If so, expect a flush of dopamine so that we continue to pursue the new activity but less so with each repetition. This paradigm creates the treadmill effect: we strive to get happier and obtain greater pleasures, but are never satisfied, leading us to pursue more. Mindfulness can break the zombie-like chase our desires lead us on to ever-higher pleasure and curtail this evolutionary vestige, enabling more worthwhile and problem-solving pursuits. Meditation cultivates mindfulness and brings about intentionality and diminishment of ego. Societies will eventually learn to value mindfulness, simultaneously leading to better knowledge generation. Mindfulness practice will also diminish the esteem of animal products, which have partly endured for their hedonic value.

CHAPTER EIGHT

The Expanse of Amazing Foods

FOOD TRADITIONS (OR LACK THEREOF)

When we ponder traditions, we may readily conjure foods associated with them. Thanksgiving evokes roasted turkey and pumpkin pie. A *Quinceañera* may bring to mind traditional Mexican fare—beans, rice, tamales, and mole. Fourth of July celebration imagines a barbecue preceding the fireworks. I was born in the US, but both my parents hail from South India. This region is known for spongy, fermented foods such as the crepe-like *dosas* and round *idlis*. Both items are typically consumed with condiments—both *sambar*, the spicy tomato soup, and chutneys: ground pastes of fruits and spices. As a kid, I despised these foods and would routinely resist my mother's attempt to get me to eat them. Despite her best efforts, she often acceded and allowed me Hot Pockets or frozen pizza. As my palate expanded and matured, I began to appreciate the South Indian fare more and proudly viewed it as a connection to my heritage. After moving to California at age thirty-one, I reconnected with a cousin, whom I hadn't seen since a trip to India when I was eight years old. Fittingly, we gorged on *dosas* with *sambar* and chutney as we attempted to recount twenty-three years of life to one another.

Even though I feel an immense pride in South Indian food, I wondered how far this tradition extends back. Tomatoes, a key ingredient in the *sambar* soup, originated from South America, likely from Peru.[283] The Aztecs cultivated the tomato fruit and introduced it to the Spanish conquistadors. The Spanish subsequently traded tomatoes to Europe. In the late 15th-century, Portuguese explorer Vasco da Gama led an expedition circling around modern South Africa, connecting Europe to India for the first time via sea.[284]

Portugal also created a new connection to introduce new foods into India and likewise reaped new foods from India, such as cinnamon and pepper. Portugal established trade posts on both the eastern and western sides of India and introduced new ingredients that quickly came to invade the indigenous cuisine—potatoes, tomatoes, chilies, okra, cashews, and peanuts, among others. Tomatoes, in particular, only entered India around the late 18th century. *Sambar*, as I know it, couldn't have started any earlier. Therefore, when I'm eating *dosas* with *sambar*, I'm not conducting a séance with my ancestors from thousands of years ago; I'm eating a dish celebrating this cool, relatively new feature, the tomato.

Indian food is not the only cuisine to benefit from the introduction of the tomato. Even though tomatoes have circulated around Europe since the 16th century, they were regarded warily because of physical similarity to the poisonous nightshade and belladonna. (They are all, in fact, in the same botanical family, the *Solanaceae*.) Tomatoes, instead, were routinely used as landscape decorations, and not consumed until much later. In Italian food, the first consumption of cooked tomato was recorded as early as 1707.[285]

The first recipe for tomato sauce dates from 1797. What most of the world envisions as Italian food—at least the tomato-based pizza topping and pasta sauces—were not a basic part of the Italian table until the last 200 years or so. It took even longer for the tomato to become established in the US. Colonists would grow, but not consume, tomatoes, again using them for decorations. Only in the early 19th century did Thomas Jefferson help popularize tomatoes throughout antebellum America with his famous Monticello garden.[286] In addition to tomatoes, Jefferson showcased okra, eggplant, and peanuts, bringing their consumption into prominence.

So, are tomatoes then the exception, with many other foods having been around much longer? Milk chocolate, the modern iteration where condensed milk is mixed with cocoa powder was created by by the Swiss inventor Daniel Peter, and is only about 150 years old.[287] Peter and Henri Nestlé merged their companies in 1879 and incorporated milk chocolate into Nestlé's offerings. Tea, that ubiquitous British tradition, only gained wide traction in the 18th century after a British spy covertly gathered intelligence detailing how to process it from China.[288] The British then grew and processed tea in India and enabled cheaper, wider access. Tiramisù, arguably the most famous dessert of Italy, was seemingly invented in the 1950s.[289]

There are even more striking, recent examples. Consider the avocado. When was the first time that you tried one? Assuming that you were not of Californian, Mexican, or Peruvian descent, it was likely only after the mid-90s. The avocado invaded the US in large part due to the passage of the North American Free Trade Agreement in 1994.[290] As a result, the unequivocally greatest culinary invention of mankind, guacamole, could shortly thereafter be found everywhere from grocery stores to Subway sandwich shops.

It's not just cuisines. Even many consumed plants are among the new hits. Kale, Brussels sprouts, broccoli, cabbage and others all derive from the same plant species, *Brassica oleracea*, more commonly known as wild cabbage.[291] Most of these vegetables only came about in the last five hundred years. Wild cabbage was selectively bred into separate **cultivars**, where each cultivar accentuated a feature of the plant for eventual harvest. For the broccoli cultivar, the wild cabbage was artificially selected for prominent stems and flowers while the luscious leaves were the objective in the kale cultivar. Brussels sprouts were routinely propped in the 1990s television that I grew up with (see *Rugrats*) as the quintessential tastes-awful-but-good-for-you food. But now, Brussels sprouts are a delicacy, flavored simply with olive oil and salt and served in restaurants that require men to wear suit jackets. Our palates did not change that quickly. A Dutch plant breeder diligently bred many variants over a couple of decades to find a delectable cultivar of Brussels sprouts that made waves and so shed its off-putting buzz.[292]

With the exceptions of a few foods (e.g., cheese, sausage, bread, milk), a large fraction of foods we routinely consume met our palates only relatively recently. To put in perspective how new many foods are, consider the history of humanity, specifically *Homo sapiens*. Most estimates suggest that humans speciated about 200 thousand years ago. Most new foods only entered the culinary stable of a given population within the last 500 years, often even more recently. This means that if we compress human history to a day, most of our foods *only entered in the final minute*. Maybe this is not that surprising. After all, many traditions, not just culinary, started in that final minute. Most modern religions (e.g., Islam, Hinduism, and Christianity) are new relative to the span of human history. I concede all this. However, I wish to make a larger point about the future of our food. If the trend is any indicator, we can expect rapid changes in our food options on the horizon.

Taking up a common thread throughout this book, when we solve one problem, we can allocate bandwidth to other problems. Worrying less about starvation means that certain populations could allocate more bandwidth to other problems, such as uncovering delectable treasures and trying to promulgate the best foods further and wider. In any critically-sized city, one can access food from a variety of cultural traditions—Ethiopian, Vietnamese, Peruvian, etc. Even ingredients have proliferated. We can buy coffee from West Africa, cheese from Switzerland, tahini from Israel. As a result, we constantly develop new food trends that may or may not endure. I recall the cronut craze and the stories of the interminable queues. When I moved back to the US from Switzerland in 2018, I noticed a new trend with poke, though if anything, I seemed to be one of the last ones in the know. I could not walk half a mile in a big city without some signage advertising a poke bowl.

THE EXPANSE OF AMAZING FOODS

We inhabit a gastronomical Enlightenment. Our food choices have never been more varied nor extensive. Any food lover would be disappointed to be transported back to any earlier period. Is this some sort of bubble? Do we expect that at some point our food choice will diminish? I doubt it. We continue to improve growing, preserving, and transporting foods. Barring some catastrophic event, such as a nuclear war or accelerated climate change, I do not expect our food choices to vanish.

If anything, the corollary that I draw from the recent explosion of food choice is that there are even *more awesome, undiscovered foods awaiting our exploration.* It's as if we're the crew leading the USS Enterprise, whose mission is to seek out and explore strange new worlds, except now the worlds are culinary or gastronomical inventions. Humanity only knows a certain number of planets now, i.e., foods and recipes. As we uncover more space/planets, our gastronomical repertoire can only expand, even if we find a few duds along the way. I term this The Expanse of Amazing Foods, or simply The Expanse.

If not apparent, I metaphorized The Expanse from food chemistry and physical transformation. We can extract components from different food substrates and combine them to generate potentially superior foods. In the example discussed before, milk proteins and fat aggregate to form cheese. Could we take proteins from something else (yeast, say) and fats from elsewhere (maybe coconuts) and create something cheese-like? Perhaps it tastes better than normal cheese, or we can improve the texture so that the outside crisps and fries when cooked on the stove. Maybe it even contains more vitamins than a bowl of spinach.

You may wonder about tradeoffs. To satisfy certain human needs or desires, food must—seemingly—either taste good (ice cream), be nutritious (a bowl of spinach), or be cheap (dried ramen). In fact, there are a variety of qualities that we value in our foods. Just off the top of my head, I can posit taste, satiation, nutrition, affordability, environmental impact, ethical impact, presence of allergens, and shelf life as relevant qualities toward a purchase and consumption decision. All foods that we currently eat are defined in terms of these parameters. To simplify things, we can consider two of them: taste and nutrition. To simplify matters even more, let's consider ice cream and spinach as occupying a point within this parameter space as visualized in **Figure 17**.

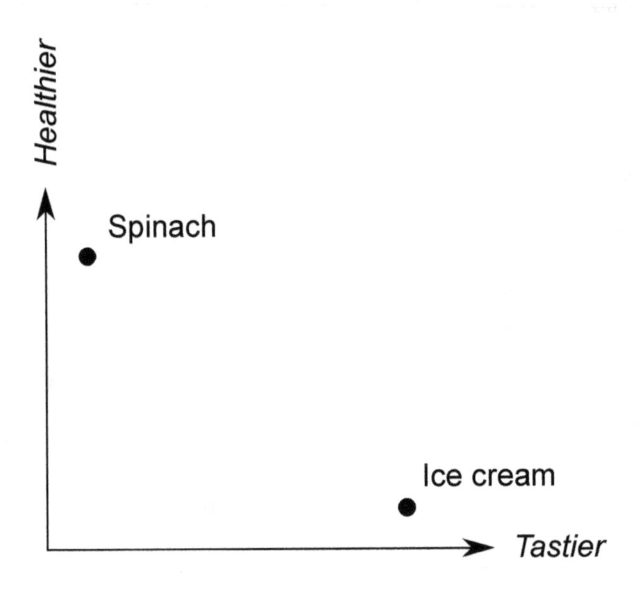

Figure 17. A parameter space of taste and healthiness. Spinach is healthier and accordingly lies higher on that axis, but lower on this taste axis. Ice cream is the exact opposite.

Obviously, I am simplifying the nutritional and taste aspect of ice cream and spinach, and the exact placement will undoubtedly vary on any individual's personal graph. Nonetheless, most of us will place them on the extremes as I have. So, let us proceed for purely instructive reasons. What about a food such as blueberries? Fresh blueberries taste delightful and are healthier than ice cream. But most people would prefer ice cream, and spinach is probably still healthier than blueberries. Nonetheless, in terms of the two parameters, nutrition and taste, blueberries clearly occupy some intermediate. Our parameter space now looks something like **Figure 18**. We can take our slipshod analysis a step further and try to evaluate which food is the best. The simplest metric I can consider is simply to add up all of the parameters into a composite score: the Amazing Index. So far, blueberries rank number one in Amazingness.

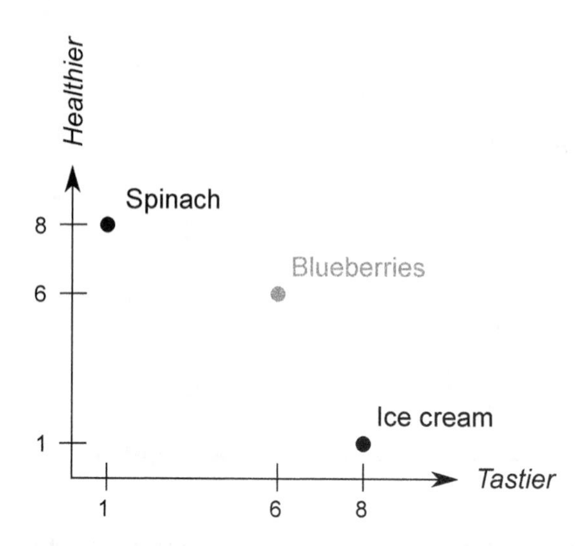

Food item	Taste		Healthiness		Amazingness
Spinach	1	+	8	=	9
Ice cream	8	+	1	=	9
Blueberries	6	+	6	=	12

Figure 18. Adding blueberries into our spinach-ice cream mix. Blueberries are the best food by our analysis when considering healthiness and taste.

Why don't more people eat blueberries? Well, blueberries have some obvious deficiencies: they're expensive and don't have a very long shelf life. So, let's consider adding a third parameter to our figure, shelf life. When we consider our Amazing Index, now the sum of three parameters (healthiness, taste, and shelf life), ice cream makes a roaring comeback here (**Figure 19**). Ice cream has effectively an indefinite shelf life given that it's frozen food and, if handled correctly, never goes bad. Over time, it certainly becomes less palatable, but it will likely never engender a foodborne illness unless there were problems in the initial packaging. So now, in the latest iteration of analysis that includes shelf life, ice cream has taken the lead, wresting it from the dainty, delicate blueberries.

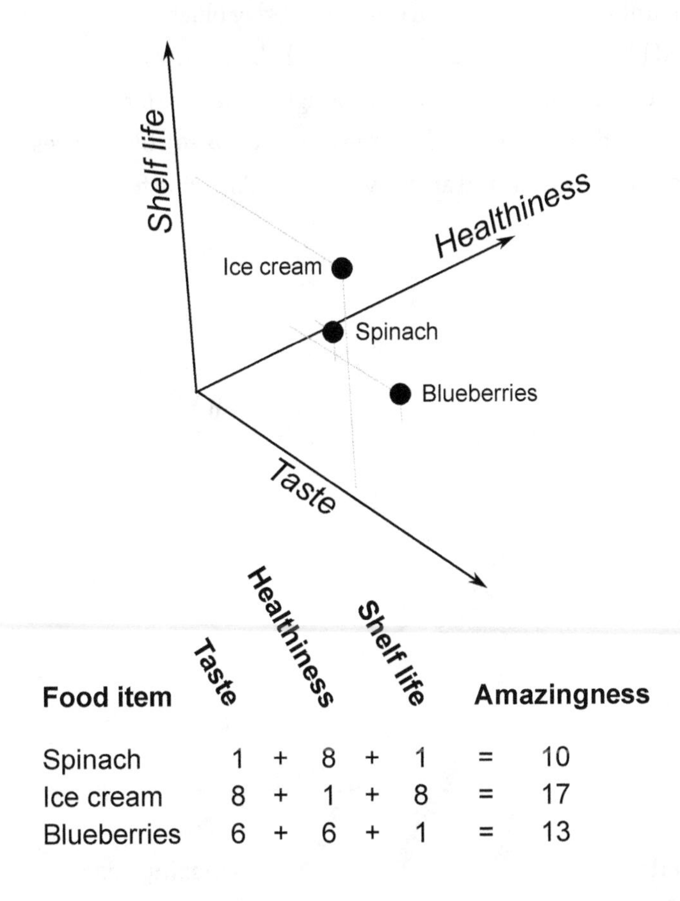

Food item	Taste		Healthiness		Shelf life		Amazingness
Spinach	1	+	8	+	1	=	10
Ice cream	8	+	1	+	8	=	17
Blueberries	6	+	6	+	1	=	13

Figure 19. Adding a third parameter. When we account for a third parameter (shelf life), ice cream overtakes blueberries as the most Amazing food.

Now, what if we could somehow improve our blueberries to improve their shelf life and affordability? Suppose we could grow them in bioreactors from blueberry stem cells. We could also employ additional modifications to make the new, improved version—the X3000 Blueberry, which tastes better, preserves longer, and is chock-full of vitamins, even more than our current-day blueberries. Our new, improved blueberries would displace the old blueberries, assuming that consumer sensibilities progress enough. When we look at the new, improved blueberry on the same plot (**Figure 20**), blueberries regain the pole position and are now the most Amazing food.

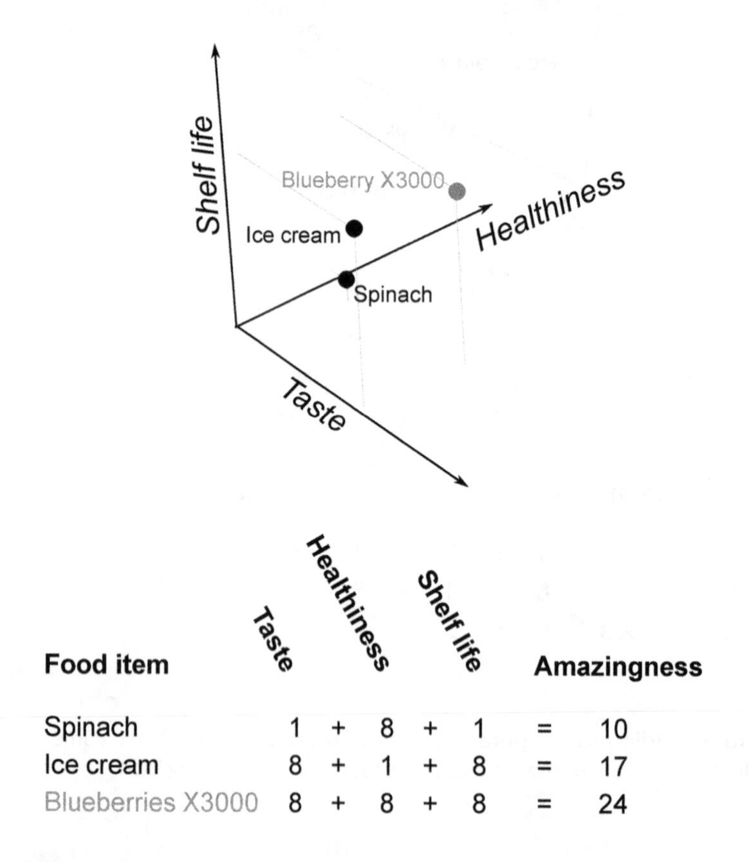

Food item	Taste		Healthiness		Shelf life		Amazingness
Spinach	1	+	8	+	1	=	10
Ice cream	8	+	1	+	8	=	17
Blueberries X3000	8	+	8	+	8	=	24

Figure 20. Pushing the boundaries with novel foods. Our innovation X3000 Blueberry takes pole position after some tinkering and engineering.

We could continue to tinker with our improved blueberry, trying to make it more nutritious, tastier, and with better shelf life, but soon we encounter physical limits. How good a food tastes is dictated by some optimal combination of various molecules (e.g., sweet, fruit, complex). There must be a sweet spot. Literally. Certainly, different people will have different sweet spots. And even the same person may have a different sweet spot at a different point in time as I discussed in the last chapter. Nonetheless, if we look at a single person at certain time, we can assume that the sweet spot is a singular point. Clearly, if we add too much or too little of the pertinent compounds, then we move away from the sweet spot. Furthermore, getting to the best possible sweet spot may imply tradeoffs in our other parameters. Perhaps adding more of the taste molecules increases the cost because we hit the physical limit of our parameter space. We physically cannot make a better blueberry. We end with the Blueberry X3000. We hire an edgy marketing firm to sell our product (hence the X3000 moniker). The blueberry physically cannot be improved more than the X3000.

However, what if we are not constrained by the Blueberry X3000 remaining a blueberry? Let us suppose that we can dispense with the pretense of blueberry-ness. What if we consider a completely different design, something completely unlike a blueberry, completely unlike any existing food? This putative food is physically possible and exceeds our Blueberry X3000 in terms of the Amazing Index. By not being constrained by having blueberry-ness, we now have more "space" to optimize this food. We can add/remove more constituents that increase the shelf life, healthiness, taste, etc. We are able to make it even better in terms of the Amazing Index, but it is clearly no longer a blueberry, and no one would make that comparison. The marketing firm suggests that we call it the X4000 and hire some Instagram and YouTube celebrities to promote it. How do we receive such a food? Hopefully, with open arms because it's merely a reconfiguration of the molecules that our body already knows; it's not anything metabolically unfamiliar.

Our USS Enterprise should gleefully seek such Amazing Foods even if they are of a completely different design, such as the non-blueberry X4000. You may think that we do not have precedents for this.

If anything, the history of technological innovation suggests the complete opposite. Consider what replaced animals in other domains such as transportation. For most of human history, horses were the default and best ways to transport humans on land. We eventually found a replacement, and it was not an animatronic horse. It was a car, a fundamentally different design. We can similarly consider a parameter space for transportation vehicles and place different technologies within. The parameters may include storage space, top speed, longevity, maintainability, etc. It is readily apparent that cars exceed horses by most parameters, enough so that the idea of using horses for routine travel is laughable today.

There are many examples of replacing animal technology in other domains. We replaced oxen with tractors. We replaced carrier pigeons with other communication technology such as telephones. Early attempts to design a flying machine resembled something that flapped like a bird. The winning design was a plane bearing little resemblance to anything in the natural world. At the airport, we still sometimes use dogs to sniff for drugs and explosives. A competing technology is a mass spectrometer that directly measures the concentrations of the compounds. In all of the aforementioned examples, the superseding technologies are of a completely different design. This discrepancy has ultimately not concerned us; therefore, I expect that we'll come around and accept the Amazingness of the unique, wholly original foods that we will eventually stumble upon.

I consider this notion to be already borne out in what and how we eat today. If someone enjoys a vegan/vegetarian dish, it is generally not because it is reminiscent of an animal product design that Impossible Burger, say, strives for. Sure, falafel satisfies some of the qualities that we seek in meat in its protein content, savoriness, and satiation. However, I don't eat falafel wraps because they remind me of lamb shawarma; I eat falafel wraps because they are delicious on their own merits. I similarly enjoy animal-free Indian and Mexican food because of the positive valence generated, not because I'm trying to recall the experience of eating chicken tikka masala or carnitas tacos. Therefore, I conjecture that we will ultimately not replace animal products with direct substitutes. Why try to make an animal-free *burger* that tastes good, is cheaper, is healthier, etc.? Why not just make *something* that tastes better, is cheaper, and is healthier? I consider this to be a much more tractable task. If we constrain our designs to something that recapitulates particular and difficult features of meat, that makes our replacement efforts much more cumbersome and slower. Not to mention, in the long run, we'll gradually supplant the animal-ness in our food, so why not go there now?

Many companies are trying to produce animal meat without the animal. They're developing *in vitro* meat technology, i.e., cultivating meat from animal stem cells. Recently, Memphis Meats received $161 million dollars in investment to do this primarily with beef.[293] I don't see this as a sensible, long-term strategy because we could readily supplant *in vitro* meat with something else that has superior metrics and tractability using microbial fermentation technology. In fact, burgers from *in vitro* technology might be a stop worth avoiding all together; it may actually be a diversion from even more

Amazing Foods. Tractability matters and begs the question, why do we need to sojourn to such a relatively inaccessible planet? We're going to have animal-free foods *that completely drown out animal products*. I envision a future where consumers are so overwhelmed by the resplendent animal-free choices that choosing an animal product would seem irrational and against one's self-interest for every conceivable reason.

The alt-food space lacks the motivation and incentives to find original foods without an animal-based analog. One of my friends worked at Impossible Foods, where they use coconut oil and potato starch in their vegan burgers. The burgers fry and crisp on the outside when you cook them. According to my friend, this was not intended behavior because a beef burger does not do this. However, this was well-received by taste testers, a feature not found in beef burgers but nonetheless pleasing. Impossible Foods for the most part is fixated on reproducing a beef burger and different cheeses as well as possible.[294] And to that end, they've actually diminished the crisping in their burgers in version two versus one. They're less inclined to travel side paths in The Expanse because of perceived consumer wariness.

This attitude is persistent in the alt-food space, and it's a shame. There are so many possibilities here. We could have a hybrid of something like chicken breast and cheese. Let's call it "chise", pronounced "chai" with "ease." Chise would have the perfect arrangement of fats and proteins, so that the outside would be chewy and crispy—think the bottom of a cheese fondue or searing cheese, such as halloumi, on a griddle. Closer to the center, chise has a protein-rich structure, like chicken breast: tender, meaty, and textured. We can flavor chise however we want. We can go for something cheese-like, add flavor molecules that grant perfect acidity, sulfur content, and nuttiness.[295] We could also go for something more meat-like with free amino acids for that umami flavor. We could do both.

There are even more opportunities. We could have coffee drinks and shakes with the perfect foam: it's more stable than a cappuccino foam, but it's not as hard and dense as a meringue. We just need to play around and find that arrangement of amino acids to get the foam with ideal airiness and stability.[296] Our bioreactor-source meat is more tender than the best filet mignon, with the smoky flavor built in, so there's no need to wrap it in bacon. We could enrich it with omega-3 fatty acids. Don't we want all of these foods? When the demand comes from consumers, the companies will happily and easily develop these products, if they see there's a market for it.

EXPLORING THE EXPANSE

Finding different ingredients is not the only way to explore The Expanse of Amazing Foods. As we discussed earlier, processes ultimately usher in everyday products, and food is no exception. Consider wine. Around 2002, Charles Shaw wine debuted in West Coast Trader Joe stores at a price point of $1.99 per bottle.[297] The affectionately known "two buck Chuck" invaded my college campus by the time I entered in 2005. I couldn't attend a party without seeing bouquets of these bottles sprouting from the drink tables.

Charles Shaw wine would never have existed without some key technologies. First, mechanical harvesting of the grapes developed to become significantly cheaper than manual processing, reducing the cost by sixty percent by 2005.[298] Second, the wine could be more automatically adjusted for acidity and sulfur content through the process of **titration**. Generally, titration is an intensely manual, parlous process where acid or base is added and mixed bit by bit, and the response variable must then be measured (pH). The desired pH generally lies on an inflection. As the person titrating nears the desired pH, they're liable to overshoot the target value. Many wineries will manually titrate their wines toward this point.[299] But the advent of automated titration reduced this labor such that a target pH is specified, and the devices add titrant, some base or acid solution, to get the wine to the desired tartness reproducibly and quickly.

Similarly, the coconut water craze crested near the end of the aughts.[300] Coconut water has long been distributed in tropical countries like Indonesia and Jamaica where coconuts grow easily. However, due to a combination of marketing and perception as a "natural" sports drink, coconut water exploded in popularity in countries such as the United States where the tree and its seed were not native. One of the primary limitations to coconut water as a mainstream grocery product is shelf life, as it is highly perishable. Coconut water teems with sugar, salts, and nutrients, a buffet for microorganisms. Being able to **pasteurize**, or remove or kill all of the microbes, while maintaining integrity was key. We have developed pasteurization techniques for milk and fruit juices, using heat applied for a short time.[301] Coconut water, due to its combination of sugars and salts, is liable to the Maillard reaction when heat pasteurization is applied. This results in an unattractive brownish product (think caramelized onions).[302] Enter new microfiltration technology in the early 2000s, which passed the coconut water through a membrane, leaving the microbes on one side and pasteurized coconut water on the other.[303]

New technologies are on the horizon, too. A friend and former classmate of mine started a company called Hazel Technologies.[304] Hazel produces packets akin to the silica gel packets that we dispose of when buying new shoes. These packets slowly release 1-Methylcyclopropene (1-MCP), a compound naturally occurring in plants, to help preserve fruit. 1-MCP is theorized to displace ethylene, the known facilitator of plant ripening.[305] Ever wonder why avocados and some leafy greens are so expensive compared to other produce? Rapid ripening is a complicit factor. Grocery stores must factor in the lost produce in the price of what they can sell. To appreciate the significance, this means that if fifty percent of avocados are unsold due to overripening, we can posit that better preservation technology would decrease that number to twenty-five percent, thereby making avocados a third cheaper.

We have food processing technologies in our homes akin to magic, too. A frying pan is like a magic wand that transforms raw ingredients into pasteurized, delicious, and digestible food. And we've never had more such wizardry than now. In the 1950s, a typical American kitchen had fewer spells to transmute food, just frying pans, oven, pots, and a toaster.[306] Today, we have blenders (both high and low power), Instant pots, pasta presses, coffee grinders, juicers, food processors, milk frothers, mixers, ice cream machines, kegerators, soda makers, temperature-precise water kettles, air fryers, countertop grills, bread machines, microwaves, and more. These devices open possibilities even unavailable in the industrial sector. For example, high-powered blenders can pulverize frozen bananas into an impressive, healthier masquerade of ice cream.[307] Blending power scales by an exponent of five with the size of the blender.[308] This means that every time you doubled the size of a blender, you would need thirty-two times more power to maintain the same speed. This is clearly unsuitable for a large scale. Altogether, our home and restaurant appliances push exploration appreciably further in The Expanse of Amazing Foods.

My own home-cooking exploration started when I entered college. After a few years, I could manage all of the foundational techniques, including sautéing, baking, boiling, and cutting. A few years later, while I was in graduate school, the molecular gastronomy revolution was taking root. For the uninitiated, molecular gastronomy is the concept of using novel chemistries and physical transformations to generate new foods such as olive oil foam or fruit juice caviar. Given my predilection for both cooking and science, I immediately latched on. I even spent over a hundred bucks on molecular gastronomy ingredients and equipment to host a molecular gastronomy party where I made fruit caviar and mojito balls.

I have not touched the molecular gastronomy ingredients since the party, and they are collecting dust in some box somewhere. Molecular gastronomy did not quite take off as I had anticipated. The foods are more of a spectacle than functional, and outside of a restaurant context, the value to that is slight. I'm never going to crave fruit caviar for myself, especially given the food waste, work, and cleanup required. Nonetheless, I do like the idea of drawing from techniques used in scientific labs for cooking, albeit, with preferably more functional impact. For example, making risotto is laborious because one must constantly stir to develop the creamy texture. In my lab, we have a heated stirrer that performs the same action automatically. It would be wonderful if I could export such a stirring device to my kitchen, but I anticipate that safety issues loom and thwart this possibility. All this to say that we still have more room for the development of new kitchen appliances, and each appliance enables easier travel through The Expanse of Amazing Foods.

WHAT ABOUT ANIMAL PRODUCTS?

You may be wondering, why not consider animal products as well. As discussed, in an earlier chapter, animals are relatively intractable as technologies. The number of transformations and explorations we can do with them is low. In the imagined journeying of our USS

Enterprise, we would have to cross a metaphorical asteroid field to reach the animal-based planet. Therefore, if we don't consider animal products, our exploration will be much faster, and we will be able to explore more of this space readily. *We will find more Amazing Foods faster if we do not consider animal products.*

I see an obvious objection to my claim. Suppose that animal technology monopolizes a certain region of The Expanse, and we have to pass through this space to find innovations in foods of certain characteristics. Such characteristics are obvious—savoriness, satiation, umami flavor—yet these properties are difficult to reproduce without animal products. Sure, we could explore *with* animal products. I see this as a potentially tenable path; however, as we discussed in Chapter 3, what separates animal products from non-animal products is better knowledge. We already see companies such as Impossible Foods and Beyond Meat employing that knowledge to create those food qualities without animals. And again, given the tractability difference between animal technology and everything else, animal technology has the initial head start though the other technologies will soon catch up and leave animal technology behind in the dust. This objection certainly may stand now, but *in the long run*, we'd be remiss to continue food innovations with animal products. The process economics of animals stressed in Chapters 3 and 4, and the intractability stressed in Chapter 5, all render animals a poor long-term investment and substrate to innovate upon.

Aside from tractability, another big problem with animal products is public health issues. In my initial foray working for a food company, I had to be trained to minimize foodborne illnesses. I completed the ServSafe Manager course, a common certification program for food industry managers for safe food handling. From the perspective of someone who does not consume meat, it's striking to see how much animal products dictate the safety procedures. Certain meats had to be cooked for minimum times at designated temperatures to minimize foodborne illnesses. The pathogenic bacteria *Salmonella*, *Shigella*, and Shiga toxin producing *E. coli* (STEC) are replete in animal products. Of course, these pathogens inhabit plants and other food as well, but for the most part, we can just wash plants and eat them raw. If I could have taken the animal-free version of the ServSafe course, it probably would have required ten percent of the time and material to learn.

What makes animal products so rife with pathogens deleterious to human health? Pathogens adapt, as with any other biological life, to a specific niche; they are evolutionarily selected to proliferate as much as possible in their choice environment. And for these pathogens, that environment is animal flesh. In animals, the Shiga toxin produced by STEC bacteria enable the pathogen to infiltrate animal cells, halting protein biosynthesis, and hijacking the machinery to benefit the pathogen's own replication. The toxin kills the cells of the animal, especially the lining of blood cells, leading to internal bleeding. Guess what are very similar to animal cells? Human cells. Shiga toxin-based infiltration can port over to our bodies without missing much of a step.

We do not have this problem with vegetables and fruits because the microbes that consume vegetables require fundamentally different machinery that does not readily degrade human/animal flesh. It's the Pareto frontier in action in that too much cost would need to be paid for the microbes to be consuming both human *and* plant flesh. In fact, anyone who eats vegetables/salad likely has such microbes within their gut, as discussed in Chapter 6. These microbes happily degrade the passing produce, and the host can even assimilate the liberated nutrients. The human and produce-degrading microbes symbiotically commune and dine together.

In similar consideration, the 2020 coronavirus pandemic is believed to have originated from wildlife farms in China.[309] Sequencing the DNA of the virus suggests that it could have evolved from a bat coronavirus.[310] Coronavirus spreads by binding the ACE2 protein receptor, which decorates the surface of our lungs. Bats also have ACE2, and the coronavirus works similarly to bind to the receptor to make entry. The ACE2 of bats and humans is similar enough that with only a few, evolutionarily tractable changes, a bat coronavirus can not only live but thrive in humans.[311] And that has certainly borne out.

It's obvious that we would have averted the pandemic if we didn't consume bats, or perhaps if China's animal farming were sufficiently regulated or, even better, shut down. Looking forward though, it's not just bats that we have to worry about. We also have to worry about avian flu, which has a mortality rate as high as sixty percent[312] though it is not as transmissible. And again, the receptors that these flu viruses bind to are similar to humans. Viruses to humans can also come from pigs and chickens.[313] So, we can understand animal agricultural activity as creating more opportunities for human pandemics. Furthermore, social distancing principles apply for non-humans, too: if animals are crowded together, then pathogens more easily transmit, replicate, and mutate into a human-transmissible form. Dense animal-agriculture practices, such as factory and battery farming, confinement, and feedlots, continue to spin the pandemic roulette wheel dangerously in favor of disaster.

On a similar note, we must also be cognizant of antibiotic interventions in animal agriculture. Estimates place seventy-three percent of all antibiotics in the world as being used for livestock.[314] Wanton use of antibiotics only increases the probability of resistant bacteria. The situation is further exacerbated because, again, animal flesh is similar enough to human flesh. It would be one thing to create super pathogens that target plants; we'll likely be impervious to them. But breeding them in animals only increases the possibility of blowback to us.

All of this actually argues for meat that is *unnatural* and original. We shouldn't want our meat to be too much like us, or animals, structurally. We should remove the ACE2 receptors or any other features that allow pathogens to thrive in our meat and in us. That's going to be much easier with protein that does not come from animals. That's the part of The Expanse that we must now explore to expediently solve these problems.

FOOD MEDIA AND CELEBRITY CHEFS

The rapid increase in the size of our Expanse is partly explained by the role of food media and celebrity chefs. After World War II, growing prosperity and access to more ingredients, (thanks in no small part to refrigeration technology)[315] increased American interest in culinary innovation. Luminary chef and Francophile Julia Child authored a seminal cookbook *Mastering the Art of French Cooking* in 1961, translating French recipes, such as soufflés, soups, and gratins, for American audiences, as well as including ingredient substitutions suited for American grocery stores. For example, using Philadelphia cream cheese instead of *petit suisse*.[316] Child earned a cooking demonstration spot on a local Boston channel that reviewed the book. The demonstration proved a hit, and three pilots were ordered for Julia Child showcasing French cooking.[317] Soon, Julia was hosting the cooking show *The French Chef*. Her show was not the first dedicated cooking show, but it marked a big advance in Expanse exploration, teaching American audiences about French cuisine. Julia is often credited, along with others, as starting the foodie movement in America.

The interest in food media increased in the following years as more shows followed, such as *Joy Chen Cooks*, *The Galloping Gourmet*, *Cooking Mexican*, and finally with a dedicated channel, The Food Network, in 1993.[318] Despite some bumps over the years, The Food Network still maintains commanding viewership, nearly the highest among non-sports, non-news cable channels.[319] This hold on cable television is impressive given the other food media in recent years. Netflix released high profile, beloved shows such as *Chef's Table*, *The Great British Baking Show*, and *Salt Fat Acid Heat*. YouTube has no shortage of popular cooking channels—think *Tasty* and *Jamie Oliver*.[320] And I haven't mentioned the number of popular cooking blogs, Instagram accounts, and websites such as Serious Eats and Minimalist Baker, to name a few.[321] The increase in food media interest suggests that we're exploring The Expanse, and ever faster over time.

Famous chefs have also helped popularize Expanse-found dishes. Gordon Ramsay's Beef Wellington, Julia Child's French Onion Soup, David Chang's Pork Bun, and Samin Nosrat's Rice Tahdig could all be trademarks, if they're not already. Some chefs are pushing us into the animal-free frontier. Dana Schultz's Minimalist Baker, with many vegan options, is already one of the most popular cooking blogs on the internet.[322] Gordan Ramsay, previously known for strident opposition to vegetarianism, introduced vegan versions of his famous dishes including the Beet Wellington and Vegan Steak.[323] The *New York Times Cooking* website now carries vegetarian and vegan recipe sections authored by famous chefs such as J. Kenji López-Alt. Vegan Chef Chloe Coscarelli captured the top prize on the Food Network's *Cupcake Wars*.[324] And Katie Higgins proved to me that chickpeas can work wonderfully in chocolate-chip cookie pies.[325] As we explore more of The Expanse in the future, I expect (and hope) for such mavens to help share the bounty.

"PLANT-BASED"

Going "**plant-based**" may seem an attractive strategy. This palatable moniker has come to replace the arousing, irksome "vegan" descriptor. Case in point: while I worked at the first vegan cheese startup, the founder emphatically vetoed that our food be labeled "vegan," but welcomed "plant-based," "paleo-friendly," or "dairy-free." Using the word "vegan" would only typecast our products for a tiny demographic and alienate a large bloc of potential consumers.

Subliminally, I take "plant-based" to mean the same thing as "vegan" when shopping for food. But looking deeply, we shouldn't fetishize plant ingredients. As argued in the nutrition chapter, the primary advantage of plant food is the fiber content, polymerized carbohydrates. While a Beyond Burger is plant-based, it's not entirely meeting the advantages of what plants confer nutritionally as there's very little fiber in it. I feel confident in proclaiming a Beyond Burger's healthiness over a beef burger, but it's not a substitute for a bowl of kale and chickpeas.

Plants are the same concatenation of chemicals and molecules as most foods, including meat or processed foods. On an ingredient label for tomato sauce, a food producer can get away with listing just "tomatoes." Beyond the amino acids, sugars, and proteins that we discussed in the nutrition chapter, plants contain numerous different compounds, including ascorbic acid, glutamate, and nitrates. The number of ingredients on a food label is utterly unhelpful, as we often see naturalists lament without an understanding of chemistry:[326] If we enumerated out the names of known molecules in a tomato, as we do in a "processed" food, we'd have a list in the thousands, if not millions.

Sticking with "plant-based" descriptors limits our exploration of the Expanse. Ultimately, the provenance of food should not matter. What matters is whether we have good explanations to highlight foods' benefits such as metabolic buffering capabilities. There is no reason that we wouldn't be able to develop plant-like foods that are even healthier. We could imagine customizing a food such that its fiber structure is more digestible by older adults and babies. Fibers would be easier to produce and tailor in microbes than plants.[327] And undoubtedly, when it comes to pure protein, microbes will win the day. The economics and tractability are just too great. Like *in vitro* meat, plant-based burgers may not be a stop to linger at for too long in our great march forward, and I implore the animal-rights movement to stop extolling them so much. It's not the best long-run strategy. Speaking of strategies that foster a future without animal products, technology and knowledge innovation can be facilitated by civic and governmental actions. So, let's discuss that next.

CHAPTER TERMS

- **titration**: adding a solution (titrant) to another food/liquid to achieve the target property. Typically referenced in regard to pH (the acidity).
- **cultivar**: when a species of a plant is bred to accentuate specific features. For example, a cultivar of a plant species may enlarge its stem. Another cultivar may draw out sweeter fruit.
- **pasteurize:** to kill or inactivate contaminating microbes in food. Pressurized steam and filtering are example techniques.
- **plant-based**: a marketing-friendly term for vegan food that, unfortunately, does some harm by placing plants too high on the food pedestal

CHAPTER SUMMARY

Despite the deep and widespread association of certain foods with certain cultures, most foods we eat today are new, only spreading in the last 500 years, if not later. Furthermore, knowledge generation and globalization enabled the spread of adoption of these ingredients and cuisines. Therefore, going beyond existing food traditions is actually unceremonious and not some profane act as there was hardly a history to begin with in most cases. The other side of the coin suggests that we're merely scratching the surface of potential new foods, in particular foods that are original and avant-garde. We could continue to make innovations with animal-based foods, too; however, this is a poor strategy in the long term. Why? First, animal foods are intractable, so any exploration will be more difficult compared to facile meat production with microbes, and animal-based products carry tremendous health risks due to the similarity of animal flesh and our own. Finally, we should not limit ourselves to plant-based foods. Sticking to merely plant ingredients also slows the transition away from animal products because it unnecessarily constrains exploration of The Expanse of Amazing Foods.

PART FOUR

Catalyzing A Future Without Animal Products

CHAPTER NINE

Realizing New Technology

INVESTING IN RESEARCH AND DEVELOPMENT FOR ANIMAL FOOD REPLACEMENTS

In order to explore The Expanse more fully and quickly, we need research and development in the relevant domains. Thankfully, we've had no shortage of innovating impressive technology in just the last fifty years with smartphones, personal computers, the internet, positron emission scanners, MRI machines, mass spectrometers, and DNA sequencers. This all begs the question: how do we practically promote the generation of new technology and knowledge? Even more, can we direct our knowledge generation efforts toward specific problems? We broached this question first in Chapter 2. We do this currently in human health research, as we pour billions of dollars into a certain problem. And we do likewise with renewable energy.

We invest in renewable energy for a number of reasons. There's the tangible issue that the sources that power non-renewable technology, by definition, will run out. There is only so much fossil fuel in our ground, and we use it at a rate faster than it can be replenished. In contrast and as the name suggests, renewable energy processes can be run indefinitely. A second reason that floats to the forefront pertains to energy security. Energy is the lifeblood of industry, and thus the engine of the economy of each country. This dependence on energy naturally requires unbroken access, a perilous situation when one must often import from another source because a country's own production is insufficient.

As of 2019, Russia and Saudi Arabia were the world's largest suppliers of fossil fuel.[328] Both countries have appalling records in terms of human rights and civil liberties,[329] yet because of their fuel reserves, they are not punished by the international community. Shifting to renewables means that countries such as South Korea and Japan aren't held hostage to such relationships for being geographically unlucky enough to possess zero fossil fuel resources. And the third and final reason is that renewable energy adoption has been promoted for environmental reasons. For example, ninety-one percent of Germans support reducing carbon dioxide emissions, and this sentiment has spurred the *Energiewende*, the planned transition from a high-fossil fuel, high carbon emission energy infrastructure to a low-carbon emission, renewable one.[330]

There's no physical reason why renewable energy cannot eventually displace traditional, fossil fuel-based technology. Similar to animal-free technology, the former could be better in every way compared to the latter, particularly with a game-changer such as nuclear fusion, which works at energies roughly a million-fold higher than current electron (oil combustion) or photon-based (solar) technologies. To this end, we've been intentional about our development of renewable energy worldwide. In 2014, humanity spent just a shade over $10 billion dollars on research and development, with the Chinese state being the largest governmental sponsor.[331] The total investment into renewables is impressive: over $250 billion dollars per year financing mostly utility-scale projects, such as wind turbines and solar power plants.

The same reasoning behind investing in renewable energy applies to finding alternative, non-animal-based foods. While animals are a renewable technology, they're a terribly costly, inefficient one, as argued throughout this book, requiring more water, land, and energy than a fermentative process for the same amount of protein production. Furthermore, animal technology isn't decoupled from the development of renewable energy. Consider renewable biofuels, which are produced from corn, the same feedstock for animals. Therefore, increased demand in animal agriculture would keep biofuel costs higher. Even if the feedstock isn't the same, plant producers must weigh growing and selling to the animal agriculture or biofuel industries.

Secondly, food security should remain a concern for many interested countries, and one way to improve food security is with alternative meats and foods. Susceptible countries with dense populations that lack arable land have grokked this argument: Singapore currently imports more than ninety percent of the food it consumes, and the government has pledged over $100 million dollars toward research and development of alternative foods, with a keen interest in non-animal meat.[332]

Finally, the environmental problems of animal agriculture are pronounced. A good chunk (over a quarter) of methane-gas emissions stems from guts of cows.[333] Methane traps thirty-two times more heat in the atmosphere than carbon dioxide.[334] And as highlighted earlier, the rapacious land capitalization of animal agriculture makes it harder to offset carbon dioxide emissions with more forests.

Even large, already agriculturally-productive countries such as China and the US would benefit from developing alternative proteins and foods. China lost thirty percent of its entire pig supply in 2019 to an African swine fever outbreak.[335] The 2020 coronavirus pandemic revealed how precarious the US meat supply chain is: Smithfield Foods shut down a slaughterhouse in Sioux Falls, South Dakota after a coronavirus outbreak among workers in early April 2020. Meat production for the entire country is concentrated at these slaughterhouses to a remarkable degree. The Sioux Falls facility itself accounts for four to five percent of the pork supply in the entire country. Its shutdown along with other processing facilities led to a spike in meat prices.[336] Animal-free technology would not suffer from the same issues as production can be more distributed, flexible, and hedged. Even though China and the US are current leaders in agriculture production, staying ahead means being at the forefront of technological innovation, lest the lead diminish. American automobile production is a good example. In the 1950s, America was the unquestioned leader, producing eighty percent of the world's automobiles.[337] However today, that figure has diminished to under ten percent due to innovations from China, Japan, and Germany.

Most governments have been oblivious and slow to react to the movement away from animal-based production. Instead, private investment into alternatives to animal products has picked up the slack and continues to grow. There's never been a better time to be a startup entering the alternative food space. Venture capital funding is vast, exceeding over $100 billion dollars in 2019.[338]

Various startup incubators and funds have leapt at the opportunity to bolster this burgeoning industry including Y Combinator,[339] Big Idea Ventures,[340] Founders Fund,[341] and IndieBio.[342] The *in vitro* meat company, Memphis Meats, raised the largest single round of private investment in the alternative food space, to the tune of $161 million dollars, in early 2020.[343] And in May 2019, Beyond Meat completed a highly successful, enriching initial public offering, where company shares became available to purchase publicly.[344] The company's total worth increased sixty percent in one day.

While the amount of funding and broadening opportunity for the alternative food industry is encouraging, it is not enough. The lack of governmental funding means fundamental research at the academic (university) level is almost nonexistent, research that could build the fundamental science to lay the foundation for applied research and commercialization as it has in other areas such as pharmaceuticals, agriculture, and computing technology. Governments, rightfully, fund the fundamental research behind renewable energy to the tune of over $10 billion per year, but not even a fraction of that money is available for alternative food and protein research, even though the arguments consistently overlap.

Right now, the only funding for university labs to perform fundamental research into alternative proteins comes from non-profits like The Good Food Institute[345] and New Harvest,[346] and only in the order of millions of dollars, not billions of dollars, or nearly a thousand-fold less per year as such organizations simply do not have the resources of a governmental organization. (Some good news in this front after I initially wrote this section: the US National Science Foundation funded a cultivated meat project in September 2020.[347]) New Harvest limits funding to cellular agriculture (*in vitro* meat) technology, which I remain unexcited about. The Good Food Institute funds alternative meat, but not alternatives to dairy or other animal products.[348] To me, the utter lack of governmental support has been a marked shortcoming in the transition away from animal products, when the outcomes for efficiency, food security, and environment are so starkly beneficial.

There are many opportunities for government funding in the alternative foods space, particularly platform technologies that would support the industry as a whole. One example is centralized standards and a database for food properties (such as texture) based on the underlying molecular structure. Many of these properties cannot be predicted from molecular details; we must instead measure and ascribe metrics (e.g., stretchability, viscosity, and brittleness) to a variety of foods. A government could set up the data standards and the repository so that researchers may share and standardize the data. Furthermore, with more data aggregated and accessible, we can start to build some models that predict certain dimensions, thus dramatically shortening the time to discovery of new foods and ingredients. The US started an effort, the Materials Genome Initiative, in 2011 to perform similar services for advanced materials, and it's been largely a great success, innovating for example a new material for the United States five cent "nickel" coin.[349] We might be able to find a new casein or new muscle mimic with a similar strategy.

The materials from such a project could be used as the ink to support another government investment opportunity in 3D food printers. The device would extrude comestible paste in a programmed mechanical fashion, forming pieces of meat. It could create the Chise discussed in the last chapter. After extrusion, the material could bet set with heat or air dried. As far as I know, no company or venture is working on such devices to sell at the consumer or restaurant level. A 3D food printer, other than self-assembly technology, is the only way that I can conceive of to have personalized food. We can have food tailored to a person's physiology with the right taste, digestibility, and nutrients. It would parallel the personalized medicine revolution, where drug doses can be dialed to each person's individual physiology. A 3D food printer would be a platform technology akin to personal computers for which many other companies can develop ink and substrates. We could even have an iStore for different recipes.

And finally, there are numerous methods for novel food discovery, which will be discussed later in the chapter. Democratizing scientific instrumentation and lab access means more bandwidth can be budgeted toward this. Enter cloud labs. Cloud labs offer researchers remote access to shared scientific instrumentation.[350] In this way, researchers do not have to purchase and maintain expensive scientific instruments. My postdoctoral tenure benefited from over a million dollars' worth of mass spectrometry equipment; none of my eventual publications could have happened without these instruments. We want similar opportunities for researchers and developers in the alt-food space especially. We see some cloud labs such as Emerald Cloud Lab (for all of life science research) and Culture (for bioreactor capabilities) are already helping in these pursuits.[351] Nonetheless, these cloud labs can struggle with funding, which I know firsthand from my experience at Emerald. Given the value and problem-solving capability, for at least alt-foods, cloud labs deserve governmental funding consideration.

I've encountered skepticism for government funding of applied efforts (like the 3D printer and cloud lab technology above). So, let's see how innovations have played out before in other industries with the help of government funding.

GOVERNMENTAL INNOVATION FUNDING

The most salient examples of technical innovations invoke emblematic companies, especially the behemoth Silicon Valley titans like Google, Tesla, Apple, Facebook, etc. It's difficult to imagine my life now without internet searching, smartphones, and two-day shipping, yet many of those technologies only arrived ten to twenty years ago. It would be easy to conclude that we need more of these companies in order to develop self-driving cars, drone delivery, and *in vitro* meat. These companies are not shy about suggesting such as well: they would happily take more tax cuts to support "transformative innovation and research."[352]

However, giving the companies the full credit does not track with reality. With a simple thought experiment, we can conclude that their business models would not have worked a century ago when there was no internet, and computers were more a theoretical construct than reality.[353] Clearly, these businesses are built on a variety of previous technologies and knowledge. We might conclude that the precursor knowledge was generated by other innovative companies. Edison's company invented the light bulb, a necessary development to run modern facilities. Alexander Bell invented the telephone, and his company's in-house lab later invented the transistor, the basis for all modern computing.

But for other antecedent technologies, it's clear that government played an indispensable role. The internet grew from a Defense Advanced Research Projects Agency (DARPA) initiative to create computer-to-computer communication between Los Angeles and Menlo Park, California in 1969.[354] Satellite technology was created by the Soviet Union with the launch of Sputnik in 1957 with purely governmental efforts.[355] Satellite technology then enabled the development of Global Positioning Technology (GPS), which started from a collaboration of the US Military and Johns Hopkins University in 1972.[356] By the late 1990s, companies such as Google had a large canvas to work upon. Their business model would have been unfeasible without all of the previously described technologies in place. Punctuating the point further, Google's core technology—internet search—was funded by the National Science Foundation as the founders Sergey Brin and Larry Page worked on it for their PhD thesis. It's fair to say that Google could have never developed, nor thrived, without direct governmental intervention and efforts.

Technologies such as the internet are special—their impact easy to appreciate and their development easy to trace. This does not always happen when seeking antecedents in precursor technology and knowledge. Fundamental knowledge and technology, as discussed in Chapter 2, ripple and cascade throughout society. When impactful knowledge and technology is discovered, many populations can benefit. For example, solving the structure of DNA in 1953 advanced the understanding of human health in foreseeable but arguably indirect ways. It led to a revolution in molecular biology, which reveled in innovation after innovation in the 1950s such as decoding the amino acid basis for proteins and uncovering the central dogma.[357] Knowing the structure of DNA helps us understand and tackle ailments such as cancer, which occurs when DNA mutates. We know that the arrangement of atoms in the DNA structure leaves it susceptible to damage from sunlight, and therefore skin cancer.[358]

This effervescent property of fundamental technology actually dissuades private companies from pursuing such scientific research *because* it is so broad and not just applicable to their narrow niche. Private companies would love more knowledge and technology, but *only* if it selectively helps them. Fundamental knowledge might help out a competitor and ultimately lead to the demise of any other company. Economist Kenneth Arrow prominently discussed this idea in 1962,[359] drawing a parallel to a landlord. Landlords will seek to maximize the rent on the buildings that they own and control. A single landlord is not going to want to pay for a public park nearby, especially when competing with neighboring buildings for tenants, even if the park would provide more benefit relative to the cost.

Amazon, Microsoft, and Google spend about ten to fifteen percent of their revenue on Research and Development.[360] Apple only spent a paltry three percent in 2014. Marketing budgets, which are clearly designed to maximize the prosperity of the company, were actually about the same, if not higher for these companies.[361] But perhaps the easiest way for a company to enrich itself without doing anything innovative is with stock buybacks. When a company has excess cash on hand, they can purchase their publicly traded stocks, driving their stock prices up higher and benefiting only their shareholders. Apple spent over twenty percent of its revenue on stock buybacks in 2018, dwarfing anything spent on Research and Development.[362] It is easy to conclude, however cynically, that any societal benefit these companies provide is incidental to their primary objective of raising as much capital versus what's invested.

The implied social contract that we have with private companies is that they will provide societal value (solve problems) while generating income for their owners: the shareholders. Their societal value comes from jobs that they provide, goods and services rendered, and the technological and knowledge generation that they then spread and monetize. This works out beautifully with internet search and smartphones, and ideally, we could steer companies toward such efforts instead of buying back stocks, which do not solve societal problems. Government regulations are a way to align the companies' primary objective (increasing returns on invested capital) to societal goals (solving problems). Limits on stock buybacks should be explored, and I'd be keen to learn the potential downsides.

We also want to spur new technology forward. I highlight a model proposed by the economist Mariana Mazzucato in her effulgent book *The Entrepreneurial State*.[363] Mazzucato argues that foundational technology is the cradle of every burgeoning industry, as the ARPANET paved the way for internet companies such as Facebook and Google. Her book covers a panoply of similar examples. Pioneering technology must be supported by large institutions with deep pockets and motivated primarily by societal improvement rather than individual enrichment. Furthermore, private industry operates on short time windows because board, investor and shareholder meetings occur quarterly, and demonstrable metrics—with implied concomitant growth—must be shown. That PowerPoint slide on earnings per share can be augmented immediately with stock buybacks rather than the same cash used for a long-term research project. Even venture capital firms which invest in startups prefer internet technology because such companies grow quickly (so-called scaling) and can quickly boost the internal rate of return metrics in geometric versus arithmetic progression compared to other industries.

In Mazzucato's and my own view, investor patience is also a key determinant to the development of innovations. Right now, the generally accepted time horizon for private companies to go without earning a profit is about three to five years. This can be excused if the company can demonstrate rapid growth, as in the case of Uber or WeWork.[364] But private investors will not tolerate toiling away on research, even if the technology in question demands such time. So, if we're relying solely on private enterprise, we'll be leaving a lot on the table. When I worked at Emerald Cloud Lab, I constantly heard about the difficulty of raising money from venture capitalists. Even though Emerald solved many problems that seemingly no other company or entity could address, the timeline was just going to take much longer. That three to five year window for complete development is not a reasonable expectation for a fully capable cloud lab. And without such a platform technology, consequent alt-food innovation slows. In general, biotechnology, including alt-food, just simply takes longer to develop compared to computing technology.

Mazzucato offers a solution to this problem of short time horizon to profitability: governments should be the long-term funder. They can patiently cultivate and foster the development of groundbreaking technology without expectation of immediate returns or song and dance performance metrics (e.g., revenue numbers). In fact, there have been successful instances of this when, in 2010 for example, the US Department of Energy provided a $465 million loan to Tesla Motors.[365] In the time period since, Tesla's total market cap, i.e. valuation, rose nearly 100-fold.[366]

The government has certainly had some failed investments too. Taxpayers lost over $500 million on a loan to the failed company Solyndra, which was touting revolutionary solar technology[367] but ultimately foundered on the shores of competition from low-cost Chinese providers. Lawmakers on the other side of the aisle were quick to castigate the loan program, but overall the program has turned a profit, and the bigger windfall is the creation of new technology that benefits society wholly.[368] Certainly, we should do our diligence to understand why such failures occur, but nonetheless, it's unreasonable to expect that the government or any other knowledge-generating entity will be capable of producing knowledge without failures. Heck, shouldn't we apply the same criteria to investment firms? Most private venture capital firms only expect a small fraction of their investments to actually yield profit.[369] The 1-in-10 winners might actually carry over ninety percent of the firm's revenue.

As I stressed in the early chapters, knowledge generation requires conjectures and leaps. Of course, we shouldn't be taking random leaps but entertain only plausible ideas by, for example, refuting bad ones with projections, back-of-the-envelope calculations, and understanding all of the potential physical limitations. However, despite our best efforts, we'll never be able to predict how new knowledge will emerge. Again, if we can figure out how a technology fares, we already know everything about that technology, and therefore, we have that technology. So, we should observe realistic expectations for how governments pursue such ventures and celebrate their entrance as filling an important void.

Speaking of innovation gaps fulfilled by government, another critical one is the "valley of death."[370] This is the time for an emerging technology when the fundamentals of the project are established and validated, and the potential of the technology is clear, but sizable funding is needed to lower the risk of commercialization.

I alluded earlier to the case of the internet where the government takes much credit. By 1969, the communication, technological, and scientific ideas behind ARPANET were well-founded and validated and even the commercial potential was apparent. What was lacking was proof of concept—a demonstration that this precursor internet could work at scale—and even the modest prototype demanded a significant amount of funding, to the tune of $25 million ($170 million today, adjusted for inflation).[371] Enter the US Department of Defense, which had the resources, patience, and desire to see if the project could benefit the military and so funded the demonstration through their Defense Advanced Research Projects Agency (DARPA). DARPA funds large projects stalled by the valley of death, in particular, generally funding demonstration-scale processes.[372]

Such funding would be valuable for the alt-food space when you consider that a complete demonstration bioreactor production-to-consumer product formation process generally costs on the order of seven figures.[373] The Quorn process discussed earlier used an unconventional bioreactor design, specifically a pressure cycle-fermenter.[374] I expect other bioreactor-derived foods to observe similar uniqueness, so it's not just a matter of making one plant for all food; valley of death funding is needed here. Likewise, a pilot plant to make 3D food printers would be similarly expensive.

In 2009, the Advanced Projects Research Agency-Energy (ARPA-E) was founded as the clean energy equivalent to the original defense-oriented agency (DARPA). For example, a promising photovoltaic for solar panels will never make it to market without a suitable production process, which is costly to develop. ARPA-E helps the inventors develop this process, providing significant funding for such promising projects (in the millions of dollars), on much more sustainable terms than they would be able to receive privately. This model has been resoundingly successful. Of the over 800 projects, 145 of them have received follow-on funding of $3.2 billion dollars from private firms versus $2.3 billion spent, and eighty-two of them have formed new companies—all of this in the span of merely a decade.[375] It goes without saying that I would be highly supportive of a similar organization for the alternative food space, and I'm not the first to suggest this.[376]

So far, I've mostly focused on applied technology. As the DNA structure example shows above, we clearly want more fundamental knowledge, too. Sometimes such knowledge cascades in unforeseen manners, as we see in the case of CRISPR-Cas9 technology, which has been lighting up the synthetic biology world for the last seven years or so. CRISPR-Cas9 is the most promising technology for powerful, directed, and facile genetic engineering to date. It's now realistic to be able to permanently eradicate genetic diseases such as Sickle Cell Anemia, which stems from a well-characterized, singular genetic defect.[377] CRISPR-Cas9 didn't appear out of thin air though. It was found thanks to curiosity-driven research into bacterial immune systems.[378] After a certain point of understanding the natural CRISPR-Cas9 system, the potential application of the technology became pronounced. The way science stumbles into new, different areas is beautiful and should be welcomed and fostered.

François Jacob, a Nobel laureate for pioneering work on bacterial physiology, couches the intentional, evidence-driven aspect of science as "day science" versus the highly creative, wandering aspect as "night science."[379] Day science is the conventional—but incomplete—picture of scientific research where a hypothesis is followed by experiment followed by conclusion. If a hypothesis is falsified, then what? Enter night science, where hypotheses are generated, i.e., the conjectures side of the Popperian model discussed in Chapter 2.

All scientific projects have elements of both day and night science and jump back and forth between the different modes.[380] However, different labs and scientists prioritize and wield day and night differently. My post-doctoral advisor, Uwe Sauer, advised me to start my first project by feeding sugar drops to starved bacteria and observing the kind of signal they elicited on a mass spectrometer. Uwe calls this process "hypothesis generation" and saw our spectrometry technology as a hypothesis generator. Once we had hypotheses, we could then test each one with follow up experiments. We shifted back to the day science modus.

The stumbling of night science took me into studying bacterial cell division, an area that I had almost zero experience with, but our data and technology had clear implications for the field, particularly fifty-year-old questions about predicting when bacteria divide based on their molecules (proteins, DNA, metabolism) inside. I eventually got in touch and collaborated with Suckjoon Jun, a physicist, biologist, and self-professed "night" scientist. His website profile reads as follows:

> *We value both logic and intuition, but more of "night science" than "day science."*[381]

Suckjoon Jun also studies bacterial cell division and developed widely influential quantitative division models.[382] These models were enabled by another physical invention: the mother machine, a microfluidic device that seals mother bacteria into chambers and captures when they split into daughter cells.[383] The videos are striking, stunning, and available on his website.[384] Our efforts traced the division event to the quantities of the FtsZ protein (as described in Appendix A), and both our labs published complementary papers on the findings.[385]

Both Uwe's and Suckjoon's night science were enabled by their internal technology development, each of which pushed the boundaries of what was explorable scientifically. For Uwe, the technology was funded by ETH Zurich, our institution's research commission. Professors at ETH can receive discretionary funding to try out night science projects and so build cutting edge technology.[386] These funding instruments contrast with the familiar ones, such as the National Institute of Health, R01, where the supplicant poses a hypothesis and a clear, detailed investigative plan, i.e. day science. There are few opportunities for explicitly night science projects along the lines of "We propose to build XYZ to generate hypotheses and follow up after that." These are effectively uncollected scientific Easter eggs that we're missing out on.

Night science, in contrast to day science, is entirely driven by curiosity. Why is curiosity-driven research so powerful? I think this, again, is explained by the best knowledge generation model.

When we're fully curious, we are at our most open: we entertain a variety of conjectures and are willing to falsify our current knowledge. The more we can do this, the better knowledge we can generate. This assertion has borne out in some real-world data. Author Dan Pink highlights this in a striking TED video about a puzzle game given to test subjects where one group is offered a monetary reward for completing the puzzle within an allotted time.[387] Another group is asked to solve the puzzle without such a potential prize. Guess which group did better on the puzzle? The one *without* the prize. The researchers offered an explanation that the group without the prize was motivated only by curiosity. The pursuit of money had an inherent tunnel-vision effect on the way the study participants were conjecturing solutions to the problem.

Therefore, I advocate night, curiosity-driven research, with an entity such as the government to patronize such efforts. Any other organizations likely have more narrowly stated missions and will fund only research perceived to be congruent to their specific mission, this versus the kind of patron we want who can recognize and appreciate the value of aimless scientific inquiry in itself. I acknowledge that curiosity-driven knowledge is a grab bag. We do not know what will come out of the research and what kind of reach it will have. I also concede that we cannot necessarily wield it to directly replace animal products. But as shown throughout the book, animal technology is fundamentally flawed, and replacement technology has the potential to take over with more development. Therefore, curiosity-driven knowledge will either do nothing in this domain or potentiate replacements of animal technology. In the US, curiosity-driven research has waned over the last thirty to forty years in proportion to our created wealth.[388] This trend is not moving in the right direction. Given that much impactful knowledge comes from curiosity, we would be remiss to not further fan that flame—and vigorously.

DIRECTED EFFORTS TO REPLACE ANIMAL PRODUCTS

There are a few research areas that could certainly use more help to facilitate a future without animal products. As mentioned, generating protein is a piece of cake. For the bulk of protein in Beyond Burgers and Impossible Burgers, for instance, the producers are unconcerned about the chemical makeup of the protein itself. The bigger concern is being able to source it easily, healthily, and cheaply, which shouldn't be a problem when using highly efficient and productive bioreactors. Therefore, finding alternative protein is not as much of a concern as other alternative food efforts. A more pressing constituent is sourcing our semi-solid fats.

The Beyond Burger, Impossible Burger, and many alternative food products use coconut oil, which fulfills many roles, as the primary fat ingredient. When refrigerated, this oil hardens so the patty isn't a wet, unwieldy mess coming out of storage. Secondly, this ensures that the fat does not seep away during transit. Further magic: when heated, say on a grill, the fats melt and form a coating around the meat that helps conduct heat, thereby crisping the patty. And finally, the fat is one of the most delectable parts of the burger.[389]

Reproducing the quality of animal fat is a more challenging task than producing protein. Animals evolved to have fat that melts at a higher temperature. For example, beef fat melts at roughly 40 ° Celsius, or just above the body temperature of a cow.[390] Evolutionarily, this makes sense as the cow would be limited if she had fat just sloshing around in her body. Imagine trying to run from a predator. Furthermore, solid fat is denser than liquid, so for storage purposes, solid fat works better. Therefore, fat synthesis in warm-blooded animal life has been evolutionarily optimized to take the form of solid fat. But the fat cannot be too solid; otherwise, it would be difficult to burn and metabolize. So, the perfect melting point is a hair above the animal's body temperature.

This characteristic incidentally also served animal flesh when rendered into meat (solid when cold, liquid when heated). Unfortunately, plants and microbes do not have the same evolutionary imperative when it comes to their fats. So, semi-solid fats are more rare among those species. However, we understand the chemistry behind fat solidification well. We've been able to take plant oils and turn them into the solid liquid chimeras, as famously shown with Crisco, through the process of **hydrogenation**.[391] This process makes liquid fats more solid by turning the doubled chemical bonds into single bonds. This in turn strengthens the interaction of fat molecules with one another, allowing them to harden and solidify at higher temperatures.

Until the mid-aughts, hydrogenation techniques were problematically scattershot, and the bonds converted to single bonds (and sometimes back to double bonds) were difficult to control precisely.[392] Even worse, healthwise, were the byproducts, specifically the trans fats. These molecules occur much less frequently in the biological world, and, because they're so chemically alien, our body has a difficult time dealing with and metabolizing them.[393] As a result, most of the food world retreated from hydrogenation back into the naturalism ideal. Premium foods such as the Impossible Burger and Beyond Burger stick with natural coconut oil to substitute for the animal fat rather than a hydrogenated vegetable oil.

However, hydrogenation technology has come a long way. The chemical company Cargill has developed a way to perform oil chemistry without introducing trans fats.[394] Additionally, the field of **metabolic engineering** has developed improved and precise ways to make target fats of interest at an industrially viable scale using, for example, yeast to grow them in a bioreactor.[395] The opportunities for creating foreign byproducts are scant when using metabolic engineering because the same enzymes are used to perform the identical chemistries as the originating plant or animal. Many of the chemical details have been worked out to engineer microbes to produce everything from runny oils to hardened waxes.[396] We should be able to produce the ideal fat, and even fine-tune the melting temperature. We could continue a step further to add nourishing fats and fat-soluble molecules. For example, we could seed our veggie burgers with omega-3 fatty acids, highly recommended for dietary intake but which mostly come from fish.[397] Metabolic engineering could create fat blends fortified with essential vitamins A and D.[398] These healthful fortifications could seamlessly blend into the fat of the food with control and precision unobtainable with animal technology.

We should also not discount other techniques to create semi-solid fats. The process of **glycerolysis** cleaves the three heads off the fat molecule triglycerides, turning them into diglycerides and monoglycerides. Interestingly, when we change the proportions so that there are fewer triglycerides and more mono- and diglycerides, the entire fat mixture will harden.[399] Enzymes from yeast can perform this chemistry[400] so we could even envision a fermentation process to produce these fats. Furthermore, we already know that our bodies can metabolize these fats, as monoglycerides and diglycerides already exist in cottonseed and rapeseed oils,[401] albeit in smaller proportions to a semi-solid version.

Another research opportunity lies with specialized proteins. While the bulk protein will form the majority mass of our foods, there are proteins with unique functions and capabilities that we'll undoubtedly want in our stable. As I mentioned in the first pages of the book, cheese falls squarely into this category. We've been able to reproduce some of the qualities of cheese reasonably well; for example, brands such as Violife get the taste nearly identical as well as some of the melting capability by playing around with levels of starch. Other properties are more difficult to replicate. Anyone who has had cheese fondue knows that cheese also has a stretchy property (which Violife cheese doesn't quite have), enabling it to wrap around a piece of bread or potato stirred into the fondue pot. This property stems directly from the casein protein, which forms spheres of fat and protein within the cheese. These casein proteins hook to one another and release constantly, enabling this stretch.[402] The uniqueness of casein has inspired many efforts to try to produce it without a cow.[403] However, as also mentioned earlier, microbes have a hard time adding the functional modifications to casein that make for the stretchiness.

Remembering the beginning of my personal casein journey, the solution that I tried was to find a casein-like protein. I theorized that given the incalculable number of possible proteins (more than the total number of atoms in the universe), we might be able to find a protein amenable for production in a bioreactor and with more desired characteristics than the original casein. In fact, others have launched some efforts validating my hypothesis. In order to find such a protein, I conjectured a **functional screen** that would enable us to search quickly for this. The functional screen works by assessing and quantifying a property of the protein.

In The Expanse chapter, we discussed how we desire certain properties of food, so why not just screen for these properties among a library of candidates? Of course, we need instrumentation and experimental techniques to do this, and that was the basis of the proposal that I wrote. I hoped to use a technique, dynamic light-scattering, to assess a protein's ability to form into the casein-like spherical structures.[404] The proposed screen wasn't perfect though because it wasn't assessing, say, the proteins' ability to stretch or to melt, like in dairy-based cheese. But these are properties that *can* be measured and screened for in proteins.[405] New instrumentation and techniques constantly hit the shelves, and our ability for screening only increases as the technology gets better and faster throughput.

Functional screening could be boosted with genetic engineering. Imagine that we find a candidate casein alternative, and we change the underlying DNA sequence to increase its potency. There would be no reason that we couldn't surpass the properties of dairy-based cheese casein. Dairy cheese casein is the product of natural evolution for a mother cow to feed her calves. My best guess is that the stretchiness and formation into a solid glob benefits the calf by being easy to consume and staying as an intact mass in the calf's stomach, a slow burn food better for the calf's metabolism. There's no reason that our unnatural food evolution can't take it a step further and yield an even better cheese fondue.

THE UNPROFITABLE REPLACEMENTS

There are problems at the fringes, too, that merit our attention. As noted earlier, the profit margins for animal agriculture are razor thin, and the industry is able to keep prices lower due to sales of byproducts. Admittedly, these make up a small fraction of the total revenue stream for producers, but if replacement efforts are tractable, then we should certainly explore them as well.

Most prominently, the hide of cows adds roughly five percent to the value of the carcass.[406] This hide is fashioned into leather goods that form our clothing, bags, shoes, and belts. Certainly, replacing all leather pales in comparison to replacing all beef products in terms of impact, but there are important considerations here. First, consumers will likely be more receptive to non-animal products that they wear versus ingest.[407] Secondly, leather is a terrible product. It wears out easily, it's not particularly water-resistant nor breathable, and tanning chemicals wreak havoc on both human and environmental health. Clothing producers already see this, and outdoor-focused brands such as Patagonia, REI, and North Face do not use leather in their products because of how crummy it is as a material. Similarly, sports-focused apparel (e.g., gym shorts) tends to use non-animal products, too—again due to performance criteria such as compression, comfort, weight, and rapid drying.[408] The Sports-focused company Under Armour is known for their synthetic polymer clothing that wicks moisture better than cotton or wool can.

Unfortunately, we are behind technological reality when it comes to dress shoes and belts, which are still predominantly leather. A few years ago, I bought my first faux-leather belt made of vinyl and polyester. Despite everyday use, it's still as good as new whereas every leather belt I've owned has worn away after a half a year or so with daily use. It's clear that we *already* have the better materials, and consumers know this: sales for athletic shoes rose 14.3% while leather plummeted 12% in 2016.[409] Similar to how achieving better food is possible *without using animals*, the same is true with fashion. The Expanse is limitless. We can have something that looks great, is supremely comfortable, lasts a while, and is waterproof. New textiles and materials will continue to render leather dress shoes obsolete—if not ridiculous—akin to what happened to fur coats. We just need more leadership from the fashion community to accelerate this trend and forsake animal products. It is conspicuously missing from icons such as Cecilie Bahnsen and Jimmy Choo. We've seen progress from the fashion industry to reduce sales of animal fur, and the logical next step is leather, wool, and its ilk.[410]

Similarly, our cosmetics teem with waste products from animal agriculture.[411] The throats, hooves, feet, and faces of cows, chickens, and pigs are boiled in acid, despite the fact that we don't know if they do anything other than change the texture of the makeup.[412] Squalene oil, used in face creams, was historically harvested from sharks, leading to overhunting.[413] Thankfully, many of these ingredients can now be made with microbial fermentation. Squalene can be produced in metabolically engineered yeast.[414] And we're thankfully seeing similar fermentation strategies for collagen production.[415] We need similar leadership from the cosmetic industry here, as the same argument applies. We can have better, kinder cosmetics if we're not shackled to animal products.

CHAPTER TERMS

- **hydrogenation:** a chemical reaction to add more hydrogen atoms to a molecule, which turns double bonds into single bonds. Hydrogenation can be used to turn liquid fat into semi-solid or solid fat.
- **metabolic engineering:** modifying the chemistry of microbes or plants to produce a useful chemical good. Yeast can be metabolically engineered to produce the malaria-fighting drug, artemisinin, for example.[416]
- **glycerolysis:** chemically turning triglycerides (fat) into mono- or diglycerides. Glycerolysis is another way other than hydrogenation to achieve semi-solid or solid fat from liquid fat.
- **functional screening:** a technique to choose and accentuate specific features among candidates. This could be selecting for candidate plants that produce the most vitamin A using an instrument that measures the amount of vitamin A.

CHAPTER SUMMARY

Worldwide, many countries have invested in renewable energy technology for reasons of sustainability, energy security and environmental concerns. These reasons apply to finding alternatives for animal products. Despite the corollary, the monetary investment for replacing animal agriculture is more than a thousand times less than what is invested in renewable energy technology. And we've seen a lack of leadership and foresight of governmental departments and organizations, which historically have stewarded some impressive technological breakthroughs, especially by sponsoring initial process demonstrations as well as patronizing fundamental science. These opportunities will not likely be pursued by private enterprises because they will be unable to reap the benefits solely for their own commercial benefit. We still have many to replace animal agriculture including trying to directly foster the qualities highlighted in the prior chapter ("The Expanse of Amazing Foods") such as with fats and specialty proteins. Finally, while replacing animal-derived food would have the largest impact, we shouldn't dismiss trying to replace other animal products in clothing and makeup. Animal-free clothing already exhibits much better properties compared to animal-based materials (such as leather) especially for the obvious-use cases of sportswear and outdoor apparel.

CHAPTER TEN

What Can Everybody Do?

OUTSIZED IMPACT

During college I was a bona fide, card-carrying Libertarian. I canvassed for Ron Paul during the 2008 election and donated to his campaign with the meager savings of a student. One of the reasons I adopted this ideology was my perception of how businesses drive progress and solve problems. In my view then, businesses could solve problems much better and faster than any other entity, and government should just get out of the way. As part of this phase, I read Ayn Rand's overlong novel *Atlas Shrugged*—all 900 numbing pages of it. Rand paints the portrait of a socialist-leaning society that's unable to solve any of its problems, including a crumbling economy and lack of technological innovation. The saviors are industrialists and entrepreneurs like its main protagonist, John Galt. In unintentionally laughable terminology, Rand characterizes Galt as a hero and the sniveling government officials as "looters" and "moochers." These heroes are blackmailed and coaxed into saving the misguided populace. Push coming to shove, John Galt and his comrades break away and form their own utopia bolstered by their individual freedom.

If it's not already obvious, this model of society and innovation is wrong, as we examined in the last chapter. Problem-solving is more complex and requires a number of actors. John Galt, Thomas Edison, Elon Musk, and Steve Jobs are not the primary determinants of technological progress and only serve one part of the whole. Aside from needing a stable society, they work using existing technology and knowledge that is often created by government or academia. There's even more to it though; problem-solving and technological innovation can be accelerated by people who aren't in the three domains of private enterprise, government, and academia. Citizens without any technical or business background can help catalyze a future without animal products.

You might ask, why you, personally, should bother? After all, many of the arguments so far conclude that this future is inevitable. Even if all you care about are resplendent gastronomical choices, then this future is better. If you care about cleaner technology that is less detrimental in terms of inputs, land, and waste, then this future is better. If you care about reducing the suffering of billions of sentient beings, then this future is better. We should want to arrive at this future sooner rather than later.

I would much rather that we dispense with animal technology sooner, and I hope that you feel the same way, too. Given that I wrote this book, I'm not content to just sit idly and let this revolution crawl into place. Rather, I hope collective actions by motivated players spur this revolution rapidly forward. The impetus is not just limited to the John Galts of the real world; many of us can affect much without a significant life change. To that end, this chapter focuses on various non-technological efforts that could speed up this transition.

CONSUMER INFORMATION

Earlier, I argued that genetically modified organism (GMO) technology would help us replace animal technology sooner, but that instead GMOs have been unfairly pilloried due to the naturalism fallacy and neophobia. Accordingly, making it easier for producers to use GMO technology will help replace animal technology, and thus, figuring out ways to promote wider acceptance of GMO technology should be a foremost thrust for animal-rights activists.

The wariness against GMOs persists not just in the US, but Europe as well. Food producers advertise the absence of GMOs prominently while anti-GMO activists have generally sought to stymie their adoption. Specifically, anti-GMO activists have sought mandatory labeling of GMOs in all food. From their vantage, it's a brilliant gambit. Labeling requirements appeal to folks who are both wary and agnostic about GMOs. Whether you support GMO's or not, or just don't care, labeling just gives everybody more information!

Of course, if there was zero cost, we should know everything about our food. Ideally, we could scan our food into an app, and we would learn everything: how the food was grown or reared, where specific ingredients were sourced, the DNA sequence of each sesame seed used, and which star each carbon atom originated from. Yes, that would be wondrous.

Earlier, I noted that a GMO descriptor is uninformative on its own because GMO is a methodology and must be evaluated on a case-by-case basis. Labeling "Contains GMO Crops" would be akin to saying "Contains Rotated Crops" or "Contains Cross-bred Crops" or even, most likely, "Contains Crops." So why single out GMO content for labeling? GMO labeling has a clear subtext because, partly, it's a strategy whereby anti-GMO activists can cast GMOs as repugnant.[417] If I smack a sticker on a jar of peanut butter that says, "Contains Hypoxanthine," the alarm-raising implication is clear. If a consumer came across two brands of peanut butter that were exactly the same except for the hypoxanthine label, then I bet the consumer would purchase the non-labeled product. In case you're wondering, despite its rankling name, hypoxanthine is indeed a real compound found virtually everywhere, especially in your food.

GMO labeling is in effect in various countries and states. The response has been counterintuitive and has, surprisingly, mollified my worries. In New Hampshire, GMO labeling diminished aversion toward GMO food.[418] A nineteen-percent reduction in opposition occurred after mandatory labeling went into effect in 2016. The study's authors did not investigate the precise mechanism behind this shift in attitude, but they conjecture that empowering consumers with more control makes a difference. Another explanation is that when we see the huge variety of foods that have innocuous GMOs, we realize our fears are misplaced. Consumers *can* accept GMO technology, a phenomenon that activists for a future without animal products should strive for and actively invest in.

Speaking of information availability, author Jonathan Safran Foer points out in the book *Eating Animals*,[419] that the animal-agriculture industry and consumers have an unspoken understanding: don't look too closely, and we'll deliver you abundant, cheap animal meat. To this end, the industry adopts cost-cutting, albeit heinous, methods. The industry itself is not inherently sadistic, merely pragmatic, trying to wring every dollar out of its process. They simply are using *everything* available given the economic pressures. If stuffing chickens into space-saving battery cages saves a dollar per chicken, even to the severe detriment of the chicken's well-being, then it makes sense to do this, especially when no one is looking nor cares.

To facilitate this willful ignorance of the reality, industry lobbyists have sponsored and enacted "Ag-Gag" laws. Ag-Gag laws in states such as Alabama and Arkansas criminalize activities such as journalists taking pictures and videos of an animal facility without the consent of the owners.[420] As a result, it's harder to blow the whistle on these facilities. It's more challenging to determine if the animal facilities are treating the animals humanely. Proponents even admit that these laws are meant to protect factory farming and that disclosure will result in negative consequences for the industry.[421] This is anti-competitive behavior that ultimately hurts both society and the way markets should function to solve problems. The spotlight on the heinous animal treatment would spur the animal-agriculture industry to adapt or to look for better—non-animal—technology, creating a win-win for both constituents and animals. Thankfully, the pro-Ag-Gag law argument has proved legally untenable. Ag-Gag laws have been defeated or ruled unconstitutional in states ranging from California to North Carolina.[422] Continued vigilance is required to ensure that these laws do not return.

Striking down these laws absolutely matters to raise animal welfare concerns. The EU passed a directive banning the use of battery cages for hens,[423] which are designed for efficiency by cramming hens into rows and columns of cages and restricting their movements. As ethicist Peter Singer notes in *Animal Liberation*, hens will often peck each other in such confinement. Accordingly, producers will often have to trim the chickens' beaks, creating another form of suffering for the helpless hens.[424] Using first-hand knowledge of the situation for animals, advocacy groups achieved similar legal breakthroughs in other countries and states such as Switzerland and California.[425]

The amplification of the importance of animal welfare opens the door wider for animal product alternatives. In the background, additional insidious, anti-market, anti-democratic machinations have been stymying the transition. The US subsidizes animal products with taxpayer dollars to a ludicrous degree that must be brought to the foreground of citizen concern just like animal welfare.

STOP SUBSIDIZING ANIMAL PRODUCTS

In the 1920s, a new era of agriculture began with the introduction of fertilizer. Farmers could suddenly produce hefty crops and flood the markets with their newfound bounty. The influx drove down prices and incentivized farmers to use more fertilizer and grow even more crops, creating a vicious feedback loop. This loop was not sustainable, particularly as it led to farmers unwittingly depleting their soil and the groundwater in such aquifers as the Ogallala.[426] Barely arable and environmentally unstable land, like the prairie grasslands, were utilized to grow crops. The precious grass that anchored the equally precious topsoil in place was replaced by less capable wheat. A drought in the 1930s decimated the crops,[427] along with their vital roots. The plants anchoring the soil down were now dead, and 100 million acres of topsoil were blown away by wind, nearly one third of what we use for agriculture today.[428] Suddenly and ironically, a food shortage arrived, and droves of families from the Great Plains moved to the fertile soils of California, the "Okie" diaspora famously depicted in John Steinbeck's seminal novel, *The Grapes of Wrath*.

To solve this crisis, contemporary president Franklin D. Roosevelt pushed the inaugural "Farm Bill" among his New Deal legislation. The Farm Bill tackled many of the issues leading to the food crisis and built a sustainable future for agriculture. The Farm Bill primarily stabilized the price of agricultural goods. The Ever-Normal Granary Program, in particular, steadied the sales and income from grain, whereby the federal government established a price for crops based on the cost to produce them. Whenever the market price for corn fell below this production cost, a farmer could take out a loan from the program with the corn as the collateral. If the price of the corn recovered, the farmer could sell the corn and pay back the loan. Otherwise, on loan expiry, the farmer could elect to repay the loan using the collateralized corn. The net effect was to render farmers less inclined to overproduce in case prices fell because the loan would steady their income until prices rose again. Thus, the treadmill effect of oversupply and low prices was averted. With the Farm Bill, the US Department of Agriculture was transformed into one of the largest departments and charged to oversee many of the country's food goals.[429] In addition to pricing stabilization, programs were instituted to educate farmers about better, more sustainable farming practices such as crop rotation and covering to intentionally protect and conserve the precious topsoil.

The Farm Bill, in terms of meeting its goals, was a resounding success. Hunger was drastically abated, farmer incomes stabilized, and warehousing of the excess food eliminated widespread food shortages in the United States. However, the Farm Bill has mutated and evolved over time, under the lobbying pressure of avaricious interests.[430] While initially the Farm Bill served individual farms and sharecroppers, farms soon consolidated into large corporate conglomerates which fast became the primary beneficiaries of the Bill's munificence. A 2014 report from the USDA indicated that only twenty-five percent of farms received subsidy payments,[431] with those payments skewed toward large farms. And from 1999 to

2012, about ten percent of recipients received three-quarters of the total payments.[432]

Furthermore, these larger farms became even more efficient, achieving records in yield and productivity year after year. The problem of oversupply and low prices intruded once again. In the 1970s, the USDA under the charge of Earl Butz posed a solution: the export and sale of US grain abroad, something which had been previously prohibited. Butz emphasized commodity crops and industrialization of agriculture versus small farms and crop variety. Butz dismantled the Ever-Normal Granary Program, and consequently, after the repeal, production only grew. This time, however, with the entire world as the market, the prices wouldn't crater. As a result, the treadmill was restarted as farms were incentivized to produce as much as possible. Larger farms were favored, wielding more political clout to maintain the unhealthy dependence on subsidies.[433]

The current iteration of the Farm Bill continues to subsidize crops, which incidentally reduces animal product prices. Between 1997 and 2005, a bushel of corn cost $2.62 to produce, but was sold at $2.00, a situation that can only work because of the subsidies of the 1996 version of the Farm Bill. Tyson, the largest producer of chicken meat in the country, saved nearly $300 million per year during this period in terms of the cost to feed their chickens. In total, industrial livestock firms are estimated to have saved $35 billion during this period.[434] In 2015, nearly $6 billion subsidized corn production of 14 billion bushels, enough to fill the entire Sydney Harbor.[435] Soybeans were subsidized to the tune of $46 billion dollars from 1995 to 2019 in the US.[436] Soybeans have two associated products—their oil and dry meal. Eighty percent of the oil is used for human consumption, but only about two percent of the meal is. The rest—a staggering ninety-eight percent—is used for animal feed.[437] Additionally, the established Dairy Producer Margin Protection Program provides direct payments to dairy producers to cover the cost differences between the price of milk and the price of the feed for the cows, to the tune of $250 million dollars in 2018.[438] Finally, the under-scrutinized Environmental Quality Incentives Program (EQIP) ostensibly helps farmers manage soil, water, and land conservation in order to meet environmental regulations.[439] EQIP funding partially helps animal agriculture stay environmentally compliant with direct funding support— another subsidy. Of the nearly $2 billion per year in the program,[440] roughly sixty percent goes toward animal agriculture.[441] Tallying all of these programs suggests that the United States subsidizes animal products by nearly thirty percent. In other words, consumers pay nearly a third less than the actual cost of producing the animal products.

This is awful. Price affects consumption. Consumers purchase less gasoline, cigarettes, and alcohol when those goods increase in price.[442] These subsidies are leading to the opposite effect; we're increasing animal product consumption compared to what would happen in a free and fair market. We can and should ween agriculture off subsidies. The agriculture industry is no longer a dainty flower in need of cosseting and diligent watering; it is vastly more technologically advanced than 100 years ago. Rescinding these subsidies does not mean that our supply chain will fall into shambles, and we'll starve to death. Analyses by think tanks, including the right-wing Heritage Foundation, rebut this fear, finding on the contrary that these industries will continue to produce and remain profitable while feeding the country even without current subsidies.[443] Furthermore, we have evidence from countries such as New Zealand that dismantling agriculture subsidies can ultimately be beneficial, leading to more diversified crops and financially sustainable farm that don't depend on government handouts.[444]

As far as I can judge, this is an issue that should strongly resonate with both liberals and conservatives who should see eye to eye and join forces to enable developing cleaner, environmentally friendly technology, reduce unnecessary government spending, and let market forces act to improve outcomes. The problem is that the argument hasn't been raised enough. And the solution does not have to be overnight nor instant. Reducing subsidies from over thirty percent to a mere twenty-five percent is a start and would help tremendously.

Imagine if we were in the gaslight era back at the beginning of the 20th century when the prospect of a commercial light bulb was imminent and obvious. The initial price differences between light bulbs and kerosene lamps did not favor light bulbs. In fact, Edison's early carbon bulb, introduced in 1883, was roughly three times more expensive than the kerosene lamps.[445] By 1900, it was cheaper than kerosene light, but not by thirty percent. Only by 1910 did light bulbs sell at more than half the cost of kerosene lamps, and by 1970, incandescent bulbs were nearly thirty times cheaper.

Now, let's imagine if kerosene lamps were subsidized by thirty percent before light bulbs were developed and came to market. How much longer would it have taken for commercial light bulbs to come to market? Consider all of the companies and individuals who would have been deterred from trying to develop light bulbs. Even though light bulbs were obviously the better long-run technology, could they have competed with financially subsidized, but inferior kerosene technology? Certainly not in the early years. How much longer would the full transition from kerosene to light bulbs have taken? If the corollary is not obvious, that's precisely where we are with animal products versus the alternatives. The subsides are a tremendous problem—we're paying to prop up bad, outdated technology and stymying the transition to better products.

If you didn't think this section could get any more depressing or you're a foreigner who remains bemused so far, well, there's more news. The European Union also flagrantly subsidizes crops and animal products.[446] The US *and* Europe also export food around the world. These subsidized grains may feed animal agriculture in other countries, or the meat is exported directly.[447] These pervasive subsidies affect animal products across the world, making it harder for everyone to transition away.

Farm Bill policies continue, I suspect, mostly out of the misunderstanding and unfamiliarity of US citizens with their complexities. The Farm Bill is generally not a hot button issue. During my research, I encountered one book geared toward a popular audience, *The Farm Bill: A Citizen's Guide* by Daniel Imhoff and Christina Badaracco. As a result, the entrenched farmers, lobbyists, and corporations can basically write bills, unfettered by the concerns of citizens. In addition, I have yet to encounter an animal rights organization tackling this issue head on despite asking around incessantly and despite how much it'd help us transition into a future without animal products. How do we expect alternatives to animal products to compete effectively in the current marketplace with such unfair advantages to entrenched entities? Consumers already complain about the price of Beyond Burgers and Impossible Burgers.[448]

The practical matter of seeing this through politically and socially is outside of my particular wheelhouse. I can highlight the problem well, but I'm a biochemical scientist and engineer, not a policy or political wonk. I do know that not nearly enough effort is being spent on such a pressing problem. Therefore, if you know of efforts or have ideas to reduce US taxpayer subsidization of animal agriculture, I'm all ears. I will muster and coordinate whatever I can in terms of resources to solve this. If it's not clear, this, in my view, is perhaps the *biggest barrier to a future without animal products proclaimed so far in this book.* We must diminish these subsidies.

INSTITUTIONAL CHANGES

In his delightful book, *The End of Animal Farming*, Jacy Reese highlights how far animal rights advocacy has progressed and assesses various strategies across different countries.[449] In particular, he extols the tractability and outcome of advocating institutional changes versus merely trying to convince consumers. Institutions such as companies, schools, organizations, and governments have different incentives though. Consider a workplace and its human resources policies. Most policies are designed to protect all of the employees and to engender a diplomatic, safe working atmosphere. Every company would rather have everyone mildly satisfied versus a situation where ninety percent of people are extremely happy and ten percent are miserable or infuriated. As a result, minority opinions and predilections can hold sway more powerfully in a workplace.

It can be easier to convince an institution to adopt animal-free practices versus any one individual. I've seen this work out firsthand in my current workplace: among the roughly eighty employees, there are only about three or four vegetarians and vegans. My company places a daily restaurant order for employees, and if a vegetarian or vegan person is working, we ensure that any restaurant can meet that person's needs. Therefore, restaurants that do not have sufficient, reasonable options may not be considered at all for our order. At the very least, we might have two separate orders, which still hurts the meat-eaters-only restaurant by splitting what could have been solely their bounty.

Furthermore, meat eaters can eat vegetarian or vegan options but not vice-versa, and they obviously can only do so when such options are available. A lot of people enjoy salad, falafel, Indian food, and Thai curries, not just the animal-free folk. All this to say that the minority vegan and vegetarian can have a more outsized impact than their proportion of the population would imply, depending on the institutions that they claim membership to. I implore folks to wield this power to promote a future without animal products. Call a restaurant ahead of time and ask what vegan options exist before ordering from them. This will force restaurants to have such options available—which is only hospitality industry best practices anyway. Implore the social and events department to have sufficient food options at gatherings. Ask for snacks that do not have animal products to be stocked.

Universities have recently experienced this effect with vociferous students. Post-secondary schools, such as colleges and universities, have more vegan and vegetarian options compared to other institutions (i.e., workplaces),[450] and many schools have dedicated dining halls for vegan students. As of 2018, two thirds of schools have a vegan dining option for all three meals of the day.[451] Even my alma mater, the University of North Carolina, has now received an A+ rating according to the system used by the People for the Ethical Treatment of Animals (PETA).[452] Many of the criteria would not have held fifteen years ago when I was there; non-dairy milk was not in every canteen.

Institutions can also be a large purchaser of animal products, so these decisions have tremendous cascading effects. An estimated seven to eight percent of milk sales go to school cafeterias.[453] This is a significant market for dairy producers, a point underscored during the 2020 coronavirus pandemic when schools closed and milk sales suffered.[454] Sales to schools have the added benefit for milk producers by indoctrinating kids into consuming dairy milk. Being reared on dairy milk means that the kid will be more likely to purchase and consume it as an adult. You might ask, well, why don't we allow non-dairy milk for school lunch? Are you ready for it? In a very real sense, this phenomenon is also part of, and due to, the Farm Bill. School lunches are subsidized by the National School Lunch Program (NSLP), a federal program instituted by the US Department of Agriculture. The NSLP defines the criteria of a nutritionally balanced lunch, which allows them to be eligible for subsidy.[455] And for opaque historical reasons, the NSLP considers dairy milk to be a nutritional mainstay. This program is vast. In 2016, the ingredients in thirty-million meals were established by this program.[456]

Unfortunately, the primary mechanism for younger students to gain access to non-dairy milk is by demonstrating lactose intolerance with a doctor's note.[457] Consider how absurd this is, that the socioeconomically-neediest children need confirmation from a doctor, who is obviously not particularly accessible to this demographic, just to *not* have dairy milk. The goal behind the NSLP is laudable: we should make sure that students, especially needy ones, meet essential nutritional demands, and what better way than through school? And cash-strapped schools are not going to turn down the easy source of money that this program provides. The core problem stems from the parochial idea of a balanced meal defined by the NSLP administrators working at the USDA. Getting the USDA to admit to the viability of nutritionally equivalent alternatives to dairy milk would be progress. I grant that children can drink water or juice instead, but at the least, getting the program to permit milk substitutes freely, i.e., without a doctor's note, would be a step forward. It's worth noting that the USDA and the dairy industry have had a historically cozy relationship,[458] highlighted by the perplexing area that dairy products garner on the overly reductive food pyramid.[459] There's also the USDA's Dairy Price Support

Program, which props up the price of dairy products by guaranteeing purchases for the school lunch programs of cheese, milk, and butter.[460] Overall, I hope that civic action and pressure on this issue will help steer the USDA in the right direction.

PERSONAL CHANGES

Let's first start with the obvious: if everyone intentionally reduced their consumption of animal products, then we would reach a future after meat sooner. Were we to do so, the sales of animal products would decrease, with many of the purchases displaced toward the alternative foods. I acknowledge that everyone may not feel comfortable nor ready to rid themselves of animal products completely. However, any bit of change absolutely matters here. There is a difference between consuming one pound versus five pounds of beef per week. Reduction is a progress-pushing, impactful strategy.

I'm not the first to make this point; in fact, there's an entire movement, spearheaded by the Reducetarian Foundation.[461] For example, followers are motivated to take up a Meatless Monday or to limit meat to once per day. Meatless Monday is a good start, and as we learn more recipes and garner more options, then that becomes Meatless March, then hopefully Meatless 2023. The feedback loop here makes it easier over time as restaurants and grocery stores adapt, followed by folks reducing even more, particularly as more and more alternatives are spurred to market, driven by demand.

It's also worth emphasizing different cultures of food. Peter Singer offers this advice at the end of *Animal Liberation*.[462] The prototypical Western meal of meat, bread, and potatoes is harder to replace one to one with an animal-free diet. But consider other cultures—Ethiopian anjera, Mexican tacos, Israeli salads, Indian curries, Thai noodles, and Chinese stir fry. These are all easy to veganize, with numerous vegan versions already abundantly available and consumed as a matter of course. Sometimes the first experience sets expectations for no meat including miso soup with tofu or chana masala with flatbread.

Normalizing veganism will also foster the future minus animal products. Right now, the US population is somewhere between two to six percent vegetarian or vegan.[463] The proportion is higher in younger demographics—a quarter of 25- to 34-year-old Americans self-report to be vegetarian or vegan.[464] There are indications that these numbers are at the bottom of an S curve, and we could see pronounced jumps in the coming years with the tipping point driven by better animal-free alternatives.[465] We've seen similar jumps for the acceptance of women's and minority rights.[466]

But in order to normalize veganism, we must societally reconcile and repudiate some outdated views. The first is just how marginalized the term "vegan" has become. It's tantamount to "pretentious" and "aggressive," and that notion apparently deters meat eaters from considering vegan ideas and food.[467] As mentioned earlier, restaurants and food producers have adopted the "plant-based" adjective partly as a palatable alternative to "vegan."[468] Secondly, some of the most uncharitable reasoning is attributed to vegans, for example that they made the choice to cultivate a feeling of moral superiority.[469] And we can't forget the meme: "How do you know when someone is vegan? Don't worry, they'll tell you."[470]

Though I'm sure there are some vegans who cultivate the feeling of superiority, in my experience, most vegans I've known have made the choice out of ethical considerations. They also would never introduce themselves with that label. Of course, it just happens to come up because of how often food intrudes into everyday conversation and social interaction—literally, breakfast, lunch, and dinner. Furthermore, it can be hard, lonely, and obviously inconvenient. No wonder we go crazy. I certainly find myself screaming inside when surrounded by obtuse, shallow-musing colleagues and friends while eating a meal together. We vegans make a choice that prevents us sometimes from getting invited to dinner parties (too complicated), restaurant outings, leaving us to sulk in a corner during barbecues or work holiday parties when no provision has been made for our dietary choices.[471] But we do it because we are concerned with the suffering and pain of animals generally. I, myself, just cannot stomach contributing to an innocent, conscious being suffering a lifetime of pain just so that I can have the most fleeting of hedonic pleasures. Sorry, but *that* motivates my dietary choices.

I know that my pain has no comparison to the real victims of the phenomena, the animals in the industry. Nonetheless, the pain of vegans is not to be diminished. Consider that just in 1988, only around twelve percent of the United States supported same-sex marriage.[472] Assuming you've adopted the progressive, sensible attitude toward same-sex marriage, imagine that you were transported back to 1988 and discussed it with folks. You'd feel frustrated, right? Or imagine that today you visited Saudi Arabia and witnessed the public stoning of a man for apostasy or a woman for adultery. And you might say that we're wrong or this is just an entitled, personal opinion, but it's more than that. We'll discuss more in the next chapter.

As part of the normalization process, we need more public veganism and more people to actually know a vegan colleague, friend, or family member, and obviously with a positive regard, too. It's not enough that Hollywood celebrities—perceived as out-of-touch and often elitist, to boot—are vegans.[473] Vegans should not segregate themselves and isolate from meat-eaters. Occasionally,

I'll see a disappointing story to that effect; for example, a vegan bride disinvites meat-eaters from her wedding,[474] and social media websites such as Reddit will lap it up because it knocks down those annoying, self-righteous vegans a peg.[475] In contrast, David Chang, a celebrity chef, was notified that a customer raised alarm that her broth wasn't vegetarian. Chang responded, "Fuck her, man. Starting tomorrow, let's put pork in every fucking dish."[476] Where is the Reddit outrage over this similarly marginalizing action?

Obviously, the bride acted rashly and even against the vegan movement she espouses, though unfortunately not enough vegans agree with me about this.[477] I deplore this kind of intentional divisiveness; meat eating is fallacious, but it occurs, ultimately, due to a lack of knowledge, opportunities, and understanding. What better opportunity to spur progress than at a celebratory wedding meal where all, including a bunch of non-vegans, are experiencing vegan food and thereby helping more animals? So, yes, please do invite them to the wedding with only vegan food. We've seen from atheism and the homosexuality rights revolution that personally knowing an atheist, gay, or lesbian person is associated with higher acceptance of those ideas.[478] I suspect the same principle to apply here. At the very least, you've spared a few animals by having others eat vegan for a meal.

The good news is that eventually veganism *will* be normalized, despite anyone's feelings or attitudes. I've emphasized throughout the book that technology will be the overriding factor. And I predict that veganism will be considered praiseworthy and make progress in the social realm as well. After all, what's cooler? Supporting the coal-era technology of food while hurting a bunch of animals *or* supporting the jetpacks of food that improve the lives of everyone? I even hope for those "No animals were harmed in the making of this" labels that we see in movies to eventually be used on our food, clothing, and drugs.

In the very first chapter, I alluded to the idea that, eventually, we'll deem using animals as a technology to be abusive and immoral in the same way that we came to collectively reject racism and sexism. Racism and sexism were wrong long before the majority of the population accepted these ideas to be so because the best morality was already clear. We're in the same position with animal agriculture: it's immoral, and we haven't seen strong refutations otherwise. We'll delve deeper into this in the next chapter.

CHAPTER SUMMARY

The transition beyond animal products is upon us; the only questions are when and how? Even though knowledge generation remains the primary determinant for this reality, many background factors affect the transition, too. Consumer acceptance of GMOs and understanding of animal agriculture welfare practices will spur this future sooner. The US and the European Union subsidize animal products to an exasperating, counterproductive degree (over one quarter of the cost) and stymie the replacers from competing effectively and fairly. Reducing and eliminating these subsidies merits pronounced effort that so far we're falling well short of. Despite currently being a small portion of the population, vegans and vegetarians can facilitate progress, most easily through institutions that are sensitive to the concerns of minorities versus individual consumers who are personally focused. Finally, vegans must canvass for their cause by emphasizing reduction of animal products generally (getting to zero eventually) and to be cherished by friends, family, citizens, and role models. Eventually, non-animal technology will be commonplace; cultivating that familiarity in a greater population is a worthwhile, productive cause.

CHAPTER ELEVEN

Morality, Animals, and Technological Progress

(FAILED) EDUCATION OF MORALITY

On the *South Park* TV show episode, "Whale Whores," the main characters Kyle and Stan join efforts to curb the Japanese hunting of dolphins and whales.[479] A roving hoard of Japanese hunters pass through aquariums, SeaWorld, and even a Miami Dolphins football game rabidly stabbing, killing, and repeatedly exclaiming "Fuck you, Dolphin! Fuck you, Whale!" In classic *South Park* comic absurdity, Kyle and Stan learn that the mouth-foaming hysteria stems from photos doctored by Americans. The photos show a dolphin and a whale flying the plane that dropped an atomic bomb on Japan during World War II. Kyle and Stan solve the crisis by providing the "real" photo—a cow and chicken were actually flying the plane. The bloodthirsty Japanese hunters now march through farms and pastures, eviscerating the cows and chickens crossing their path. In the final scene, Stan's dullard dad, Randy, exclaims with misplaced enthusiasm, "Great job, son. Now the Japanese are normal, just like us."

I'm not sure if the writers of the show intended to make this point, but they underscore an obvious dissonance. Why is hunting whales and dolphins wrong, but farming and killing cows and chickens okay? It seems like slaughtering whales, dolphins, cows, and chickens should all categorically be ethically permissible or not. Similarly, how do we know whether it is okay to eat insects or other human races? We obviously need some means to determine this; we require a morality.

The teaching of morality is not part of most of American formal, secular education, and most of us will not take an ethics or morality class unless we pursue a liberal arts college education. Wide swaths of the population receive minimal exposure to the ideas. Researcher Jonathan Haidt notes that most people draw their moral knowledge from peers and societal norms versus more thoughtful means.[480]

As I'll argue in the chapter, developing morality is no different than developing awareness and expertise in any other area of scientific knowledge. *Development of morality is a science*, and accordingly, we must yield to the best morality. I often see this blind spot in scientifically minded groups and individuals. The German-based *Kurzgesagt*, a group that focuses on scientific engagement with the public, covers topics from quantum computing to genetically modified organisms using exquisitely animated videos. *Kurzgesagt* produced a video on eating meat.[481] Generally, the video highlighted the problems well and showcased the potential alternatives (e.g., *in vitro* meat). Disappointingly, though, the video concludes that it's okay to still eat meat because it tastes good and because we will replace it soon, totally ignoring what the best morality has to say on the topic. Thankfully, *Kurzgesagt* did redeem itself in a consequent video about milk; they admirably did not shy away from the harms of the dairy industry to cows.[482]

Scientists have been quick to decisively support the best science on charged issues such as climate change and vaccines.[483] As soon as the banners are raised, scientists rally behind their climate or immunology comrades. Even if a scientist is not in the field, the sentiment is that we should always listen to the best science, and I completely agree. We even see scientists pick up the mantle for moral causes such as anti-racism and anti-misogyny.[484] If the morality behind evolving beyond animals-as-food technology is clear, shouldn't scientists raise their pitchforks here, too?

The primary problem I suspect again is a failure of moral education, even though the morality behind animal technology is familiar and well-trodden intellectual ground, most notably in Peter Singer's seminal book, *Animal Liberation*.[485] In fact, reading an essay by Singer in college inspired my growing concern for animal welfare. Other authors include Magnus Vinding and his freely available *Why We Should Go Vegan*.[486] Nevertheless, I will add more color, especially in juxtaposition to the development of knowledge and technology. Before we can go too far though, we have to answer some meta-questions. What is morality? And where do morals come from?

THE CENTRAL MORAL ARGUMENT

In the first chapter, I claimed that the current purpose of humanity and our endeavors is to solve problems. Like knowledge, the problems are conjectured and refuted and also can be of different magnitudes and scope. So, figuring out what I'm going to wear today is not the same problem as figuring out how to deter nuclear war. It's easy to explain why nuclear annihilation is more pressing than my sartorial indecision. For the rest of the chapter, when I refer to problems, I'm concerned with those of the largest, possible scope. For example, if I purposely do my taxes incorrectly to avoid paying too much, then that wouldn't be a problem for me (assuming I don't get caught), but it would be a problem for the population at large. My unpaid taxes might have funded schools, police, and public services; therefore, me not paying my taxes is ultimately a problem. That's what I mean by the largest possible scope. The scope includes future generations of people and animals. So, polluting today would be a problem for beings of the future.

I define **morality** in relation to problems. More moral behavior reduces problems, or solves or ameliorates them, while immoral behavior creates and exacerbates problems. When we consider a problem situation, i.e., if we see a child drowning in a lake, we consider all of our possible actions. We could jump in and save the child, or we could walk on. Each potential action would have associated problems: we jump in, and our clothes get wet. Or we walk on, and the child dies. We should perform the action that solves the highest magnitude of problems, in this case saving the child.

Many situations are not as cut and dried as a child drowning and will elicit controversy. Is it okay for cousins to marry and have children? Is it okay for people to commit suicide? At what stage of fetal development is abortion acceptable? We might not always have clear answers at any given moment, but we might learn more and develop better arguments, such as, a fetus develops consciousness at twelve weeks, or children sired by cousins have no significant increase in birth defects. As a result, morality changes over time as we generate knowledge and develop arguments. And the prevailing, accepted morality in society may then follow. For example, gender equality became more morally accepted and will stay accepted because any argument against it is feckless.

The idea that we can just reason our way to better morality may create uneasiness. If your amygdala is disagreeing with the above statements, I suggest the following. First, propose where notions of morality come from instead. Then ask whether that alternative is better than using the Popperian model (conjectures/refutations). Secondly, explain why moral knowledge is generated differently from any other scientific knowledge. My assertion is the most parsimonious one: morality is just like any other knowledge and the product of conjectures and refutations. Why should we have an exception for moral knowledge?

Given that morality is like other knowledge, morality is absolute and *not* **relative**, as often claimed.[487] Different morality across cultures, nationalities, and religions is not privileged or excused. The best moral knowledge stands out and informs in the same way as the best biology, physics, and mathematics knowledge. Would you ride in an airplane created by someone who invented their own physics? In the same way, we should accede to the best morality. Throwing homosexuals off roofs should be wrong in both Western democracies and Islamic theocracies. Killing widowed, older women should be wrong whether in Indian or Tanzanian rituals.[488] Morality does not come from religion or religious texts. The Bible, Quran, and Torah permit or extol actions such as rape, slavery, genocide, and homophobia. Because of the lack of Popperian error correction, these texts are outdated knowledge and no longer serve as useful moral guides.

Diversity of thinking is absolutely important. Generating the widest set of conjectures only helps us to solve problems and make progress. But some ideas are just simply awful, and we should not hesitate to affirm that, even if those ideas are cherished or endorsed by a specific group of people, religion, or culture. Slavery was morally wrong in the 19th century United States, and is wrong in present-day Qatar.[489]

Now, according to the model of morality I am presenting (solving problems), we can also conjecture what are problems and problem situations. Furthermore, we can develop a sense of what are problems to look for and to reduce. Going back to nuclear annihilation, what specifically would make it so awful? It would presumably kill a lot of people who would not want to die. It would create sorrow in their loved ones. It would create a less habitable Earth. For many of these problems, we noticed a constant theme, the problem of **suffering**. The nuclear bomb victims would suffer as they are vaporized. A less habitable Earth presumably means that the survivors would suffer. People would also suffer because their loved ones would die.

To further appreciate the problem of suffering, consider the following. If we find out a child has been kidnapped, we're heart-stricken by the suffering that child experiences. If we learn instead that the child has been kidnapped from abusive parents and is now in loving, supportive care then we see a problem as being solved, not created. We are dismayed by entity-in-question-suffering. Suffering is what's bad about the situation, not just the act of kidnapping, and it's a problem that we can generally agree upon. While other problems absolutely exist and merit our attention, suffering sticks out as a problem in many situations we can conjure (e.g., disease, pain). But perhaps not all suffering is bad. It can even sometimes be beneficial in the long run; for example, we suffer schooling as a necessary bridge to cross when learning a new topic, and to grant us humility. So to qualify this, we want to *minimize unnecessary suffering*. As a useful, illustrative exercise, try to come up with various problems worth tackling by society. Where does reduction of unnecessary suffering rank? I anticipate at the top.

So, we must now think about animal suffering and whether it's a worthy cause to tackle, but first it's worth asking the even more basic question, do animals suffer? The morality behind animal technology is complicated by the fact that humans cannot talk with animals. Women can draw and speak from their experience against misogyny. Likewise, this is true for minorities from different races and sexualities and their own persecution. Though they may communicate it in many other and equally persuasive ways, animals cannot vocalize to us about any pain or suffering they feel, and only advocates such as myself can speak on their behalf. But this doesn't change the argument, ultimately, because it doesn't matter who says it: me, a cow, or a scribble on a napkin found on the side of the road. This lack of communication ability often leads to the absurd conclusion that animals are incapable of pain and suffering.[490] Philosopher Descartes famously proclaimed that animals are unconscious.[491] But consider how absurd this idea is.

Earlier, I described consciousness as a mechanism to evaluate a number of sensory and internal inputs. And critically, consciousness is needed for species that naturally encounter many different situations, most notably generalist species such as humans. Many animals fit this bill as well. They must be able to find food, determine who is a friend or foe, deem a place suitable for sleep, etc. They need an appreciable consciousness.

Now, let's consider the purpose of pain. Clearly, pain is a way to correct our behavior and actions and avoid adverse outcomes given a certain consciousness input. Let's *not* jump into that fire and burn away our flesh. Animals need the same tooling for the same reason. Animals clearly feel pain; it would not make sense evolutionarily, otherwise. But unfortunately, that's not what we see in "scientific" studies. Many studies emphasize that because we can't get into the minds and consciousness of animals, we cannot state categorically whether or not they feel pain.[492] Therefore, let's just throw our hands up in the air and not do anything. Again, this is not science. We technically do not have a "scientific" way to access the consciousness of people as I wrote in Chapter 7,[493] so do we say that we don't know if other people feel pain or not?

I hope the folly of this specious thinking is obvious. To repeat and summarize our best epistemology: we start with a conjecture (e.g., "animals feel pain") and try to falsify it versus competing conjectures ("animals do not feel pain"). What remains is our best science. Chickens screaming after being debeaked, or calves crying out for their separated mother all support the first conjecture versus the second.[494] The most durable explanation is that animals have consciousness and feel pain. When a steer reels and screams after being fire-branded, he's in pain. Otherwise, one must present a more falsifiable explanation that also explains the observation. We must guide our morality accordingly.

In the background, we should also consider whether we're motivated to *want* to believe that animals don't feel pain. As claimed earlier, revealed knowledge is sometimes inconvenient and remains germane no matter our ego or feelings. Revealed knowledge can't care about anything. Eating bacon cheeseburgers and performing ill-conceived animal experimentation—both are certainly easier if we're convinced that rats, pigs, and cows are unfeeling automatons. Acknowledging these feelings and how they might weigh in as absurd conjectures will help refute them.

Pain and consciousness are undoubtedly intertwined. Given that consciousness is the capacity to assess sensory inputs, more consciousness should be tantamount to greater capacity to feel pain. Putting our hand on the hot stove should interrupt our consciousness in a way that drowns out anything else. Obviously, it's an imperfect heuristic, but a level of consciousness should accord with the capacity to feel pain. I would love a more precise way to gauge this, but I'm going with the best I've come across. We'll discuss this more in a bit. Prolonged pain is what I consider to be suffering. We suffer when a loved one dies or when we're worried about where our next meal is coming from.

This brings us to the central argument of this chapter. Animal agriculture creates tremendous, unnecessary suffering in conscious, pain-feeling animals. As stressed throughout the book, we do not need animal technology, and ultimately, we'll do better in the long run without it. There is nothing special or specific to it that we won't be able to reap better from other technology, whether in terms of efficiency, taste, or health. And we have much to gain, morality-wise, when doing away with animal agriculture. *Therefore, we should strive to minimize and reproach animal agriculture.* And in the next section, I'll address the tripwires that might have been triggered in your head.

RESPONDING TO POTENTIAL REJOINDERS

First, you might be asking, what about the suffering of insects, plants, and microbes? Isn't the pain of anything living argument a slippery slope problem. In other words, if animal agriculture is bad, then isn't plant agriculture bad for the same reason? Suffering is suffering. If we have strong conjectures and explanations that plants, insects, and microbes can experience profound suffering, then we must weigh those ideas in. But as stated earlier, a reasonable heuristic for level of consciousness is the quantity of neuron cells the organism possesses or the number of environments they inhabit. Plants are sedentary, and consciousness would not benefit them as much as a land-roaming animal. Microbes likewise do not have to make complex decisions about their environments. Insects clearly straddle the line between animals and plants. By the Pareto frontier described in the Chapter 4, it would be too evolutionarily costly to develop a sophisticated consciousness in these organisms.

Furthermore, in the event that we do find that plants suffer tremendously from agricultural practices, it would still morally favor non-animal agriculture. After all, what do animals eat? And given how inefficient animal technology is, if anything, plant-suffering damns the technology even more. We're causing more plant suffering by sticking with animal technology.

Secondly, we unfortunately must draw a line of moral consideration; we must make an uncomfortable choice. We just do not have the technology and knowledge to solve all of these problems in a satisfactory way. This lack of satisfaction may indeed deter change, but it's not an argument for inaction, even if it may shift over time as a function of technology. Consider if we developed chemical transmutation and could apparate food out of the air, or if we could load our consciousness onto silicon-based computers powered by sunlight. That would completely skirt the moral dilemmas presented here.

In fact, we already societally do draw a line for the life that's permitted to be farmed and consumed. For example, we clearly forbid people to eat other people. Throughout history, cannibalism was accepted under specific circumstances by a few different cultures. For example, the Wari tribe of the Amazon consumed flesh of deceased loved ones as part of the mourning process.[495] Today, cannibalism is broadly *verboten*, especially the idea of enslaving, breeding, and cultivating humans for meat, skin, or cheese. In the West, we also strongly frown on the consumption of dog and cat meat. Dogs and cats have endeared themselves to us as pets. This is not the case in many Asian societies, where both have been consumed in recent history, if not currently.[496]

More efforts have been made to expand the human, dog, and cat side of the "do not eat" line. Former NBA star Yao Ming has helped reduce consumption of shark fins.[497] And activists have reined in commercial whale-hunting practices throughout the world.[498] But weirdly, most people do not have similar feelings about cows, pigs, and chickens, which remain firmly and broadly on the other side, as illustrated in the *South Park* episode described earlier.

What we collectively consider normal is a product of humanity's inability to reason morally and effectively. Social psychologist Jonathan Haidt has documented this effect profoundly, establishing that generally people first intuit about whether something is right or wrong and then apply *post hoc* reasoning to justify their decision.[499] They take cues from those around them. What's ingrained as normal becomes a moral blind spot. This is why, historically, Americans could condone slavery, segregation, misogyny, and homophobia, and why Germans could shy away from too deeply contemplating the abduction, forced labor, and eventual extermination of Jews. The antidote is reasoning and arguing about what is truly moral, not being satisfied with our intuitions or surroundings.

If we start from the principle of trying to minimize unnecessary suffering, then that suggests we need to account for level of consciousness, which then becomes the boundary. If we as a society believe that cats, dogs, dolphins, and whales are worthy and deserve protection, we must consider other animals with equivalent or higher levels of consciousness. Cows, pigs, and octopi clearly fit the profile, and we should thus absolutely arrogate their right to avoid suffering needlessly given the members already in our circle of consideration. Certainly, chickens may be controversial and straddle the line, but I think most of us would want birdlife on our side of the "don't-eat-me-I'm-conscious" fence. The line blurs even more for fish, insects, shrimp, and oysters. And I do not have an obvious, clear explanation for the inclusion of fish, but not insects. Nonetheless, any line that we draw using this reasoning is argumentatively stronger than the prevailing one in Western society, and that's enough to warrant inclusion of more animals.

You might object that the consciousness line "feels" wrong. So, I challenge you to come up with your own line. If you do come up with something better, let me know, and I'll adopt it. Many have tried other lines, and commentators have suggested features such as species, i.e., our moral imperative is to maximize the well-being and proliferation of our species and the ones that we have relationships with (e.g., dogs and cats).

But why is species a meaningful moral attribute? Consider the concept of species historically. It's to help draw lines for biological classification, specifically with respect to biological evolution. Members within the same species can interbreed with each other but not those outside.[500] That doesn't sound like it should have anything to do with morality and the problems of unnecessary suffering. To drive this point home further, consider that we could revive *Homo neanderthalensis*, more commonly known as Neanderthals. Synthetic biologist George Church muses that this could be done for a modest sum of money.[501] And as technology develops, this will become easier.

Imagine that we bring them back. They are similar to us *Homo sapiens* but with conspicuous differences: being stockier on average, with bigger noses and smaller chins. Mentally, they are not as developed as us, but they're able to communicate and form tribes. They have culture, aspirations, and can learn new knowledge. As best as we're able to tell, they have consciousness and are able to suffer. Even if they are less intelligent on average, we'll obviously see that they deserve rights. We would quickly realize that it's absolutely wrong to enslave them, harvest them for organs, or to inflict harmful experimentation on them.

We've seen this moral reasoning play out in our media such as the book *Never Let Me Go* and the movie *Blade Runner 2049*, where the rights of "subhumans" are addressed and advocated. We can only conclude that drawing moral lines by species holds no more sense than lines drawn using race, nationality, or gender. The victims in question don't even have to be corporeal. The television show *Black Mirror* routinely harps on the issue of conscious beings who are purely digital, only existing as a virtual program in a computer. As the show emphasizes through some grueling examples, it would still be wrong to torture and, otherwise, impoverish them. (Further elaboration: I'm not arguing that "subhumans" are equal, i.e., should have the right to vote. I'm arguing that they, at least, deserve the right to not needlessly suffer.)

The value of existence is another argument that exponents invoke to support animal agriculture: all of these cows, chickens, and pigs would never have existed without our industrial practices. There are nearly 24 billion chickens in the world right now; this is orders of magnitude more than there ever were in nature.[502] Based on the value of existence alone argument, this is a good thing, right? Cows should be happy because we're creating more of their brethren, and their "species" is winning the evolution game, at least in terms of reproduction! Bummer about the living conditions and eventual slaughter, though.

But if sole existence was a positive attribute, then why don't we do the same with humans? Every nubile woman should be pregnant, and we should actively be seeking ways to make artificial wombs to crank out as many humans as possible. But we don't do this, and no one is advocating for it because we, at least implicitly, acknowledge that existence is best coupled with a life worth living.

Two partners may just have two children because seven children would diminish everyone's well-being. The same principle holds with animals. Bringing 20 billion chickens into the world to waste away in cages for their entire life does not bring value. It creates more problems than it solves. (Note that there is a difference between life that already exists versus life that *could* come into existence. For already existing lifeforms, taking away their life has at least the additional problems of their desire to stay alive and to suffer at the prospect of death. Such a problem cannot exist by definition for life that does not exist yet.)

This brings me to the final point about whether to consider vegetarianism versus veganism. I grant the good intentions of vegetarian thinking in that what's wrong with animal agriculture is the slaughter of the animals. Yes, that's a problem; however, should we want the creation of such wretched lives in the first place? Would it be that much better if heifers weren't killed, merely injected with hormones and separated from their families for their entire lives? Furthermore, products such as dairy milk lead to slaughter. Heifers must continually be pregnant in order to produce milk, and sometimes they birth a male calf, effectively a waste product to the dairy industry. These calves are turned into cheap, dairy-subsidizing veal steaks through a truly heinous process, details of which I'll spare you here but can be found online.[503]

Finally, you might ask about free-range animals or if we could ensure that the animals have a good life while consuming their eggs, milk, and, eventually, their meat. I will probably surprise and infuriate a lot of vegans here, but I see a plausible scenario where this is morally permissible if not morally good. Practically though, it won't scale. Industrial agriculture is designed to hit demand numbers for animal products such that ninety-nine percent of slaughtered animals come from factory farms in the

US.[504] As stated throughout, these ignominious practices are the outcome of the physically-limited performance metrics of animal technology and financial incentives, not some inherent sadism in producers. We'd somehow need to convince or mandate for vastly lower consumption, 10 to 100 times lower based on the numbers I presented earlier. So, going from eating meat once a day to once a month, that's a big ask of society. All of this can be completely side-stepped with better non-animal technology, so shouldn't we just go there instead?

WILDLIFE SUFFERING AND FUTURE PROBLEMS

If animals have the right to not needlessly suffer, why fixate on just animals in agriculture. Why not consider all animals, including those in the wild? For example, predator relationships have a well-documented effect on the stress levels of prey. Elk in Yellowstone National Park experience chronic terror from roving wolves, confirmed by elevated cortisol levels, the neurochemical signature of stress.[505] These elk eat and reproduce less.[506] Wild chickadee birds develop post-traumatic stress disorder after exposure to predators, with a sensitive fight-or-flight alarm when signs of another predator are present.[507]

It's not just distresses over predators. Primates who fall too low in social standing among their troop experience depression.[508] Elephants, cats, and baboons will grieve over the loss of loved ones.[509] Animals in the wild manage hunger, drought, disease, and injury on a day-to-day basis. Wild animals clearly suffer; furthermore, they lack the tools, capabilities, and knowledge to improve their station. Generations of these animals will experience similar problems without circumstances ever improving from their predecessors. And fundamentally, with very few exceptions, animals have limited capability to pass and use knowledge to solve their problems in the way that humans do. However, this biological deficit and environmental unluckiness are still not reasons to consign them to their wretched lives.

So, the best argument right now is that for whatever line we draw, wild animals must be considered as well. At face value, this seems ludicrous, and often wildlife suffering is invoked to draw the argument for veganism *ad absurdum*. However, cutting through any knee-jerk incredulity, we must ask if this is a reasonable concern. Does wildlife suffering merit applying problem-solving bandwidth from humanity? If it is truly absurd, then we should be able to shoo it away with argumentation. (The *ad absurdum* argument is also inadequate because two wrongs don't make a right. Animal agriculture occurs independent of what happens in the wild. Animals are created for agriculture that wouldn't have existed otherwise.)

Harking back to the first chapter, problems must always be tackled in stages. For example: first our immediate starvation and drought, then long-term food and water security, followed by devastating but treatable diseases (e.g., malaria and worm infections), and only then long-term, difficult-to-solve diseases such as cancer and Alzheimer's disease. Knowledge such as disease-resistant crops and vaccines potentiate our ability to handle and solve these problems.

Imagine the continued providence of new technology—solving problems only becomes easier and less costly. We solve problems such as poverty by, say, instituting a world-wide universal basic income program because of the vast wealth we've accumulated. We've reduced crime to near zero because we've learned how to cultivate and reciprocate social and institutional trust by reducing corruption, wealth inequality, and improving how justice is dispensed. We have fully personalized health care that, on a daily basis, helps individuals avert or delay cancer. When the cancer does hit, we can use genetic engineering to mobilize a person's immune cells against it effectively. In this future, we're only grappling with shallower, but nonetheless vexing problems such as figuring out how to fold a fitted sheet. The remaining human-specific problems are existent, but pale in comparison to the problems of today.

Many books have trumpeted the knowledge-induced progress of today, including Steven Pinker's *The Better Angels of Our Nature* and Hans Rosling's *Factfulness*.[510] Nonetheless, there is some balking at the idea that we're making progress. But there's never been a better time to be alive in most parts of the world, as measured by poverty reduction, safety, and access to healthcare, food, education, and wealth. Messenger Steven Pinker has received a lot of flak.[511] Given the amount of ink spent on the topic, it'd be superfluous to make the same points as Pinker, but a few comments:

- Remember, how you feel about the present is not an argument. This point is especially important because, as we saw earlier in Chapter 7, we're not naturally designed to be contented or complacent due to our evolutionary programming. News media capitalizes on this with emphasis on negative stories more intended to draw our attention rather than inform us about reality.
- Progress does not mean the problem is eliminated. Yes, racism still exists, but an elected black president for two terms is very different from the reality of 100 years ago.
- Progress occurs at different speeds in different domains; while income has risen overall, it's unfortunately not even across levels or demographics.

- Yes, some areas have gotten worse, for example, climate change. But overall big picture, would you rather live in the world a century ago?
- Even if you disagree with Pinker's fundamental thesis and claim that we're not making progress, then what's the alternative? Should we just keep things as they are and mope? If we're not making progress, then shouldn't we at least try?

Presume that we can and do make progress, then imagine it proceeding further and further. We will reach a point where human problems are trifling. In such a scenario, wouldn't it be absurd to have a group of prosperous, contented beings (humans) surrounded by miserable, abject ones (animals in the wild)? Furthermore, it's only going to become easier to help those animals. A vaccine-on-demand system would help animals just as much as it helps humans. Imagine a few button presses and dollars and we have a vaccine that curtails a devastating disease for a herd of deer. Shouldn't we use it? Similarly, once our bioreactors are producing so much alternative meat, far more than our population can and should eat, then shouldn't we share it with the wolves, who then leave the hares and elk alone?

The most common objection is that by intervening, we're disrupting the ecosystems, implying that this is a bad thing. That kind of thinking is just another misplaced fetishization of the natural world that I inveighed against in the first chapter. Certainly, we want to be careful; we don't want to make the situation worse than it is. For example, if we feed wolves and help the deer, then maybe the deer population will explode leading to downstream problems like the overconsumption of resources and thus starvation. We might want some form of birth control for the deer. So, we'll want to try experiments and learn what works and doesn't. The point is that we *should* try. We should try to reduce wildlife animal suffering and produce the knowledge to figure out the best way to do so.

To drive this point home, we societally want to care for disabled people and orphaned children. These people lack the wherewithal or ability to improve their circumstances on their own. We would even be contented to financially support them and maintain their livelihood with housing, education, healthcare, food, and socialization. The same reasoning applies for sentient animals: they lack the means to improve their situation, so let us step in and provide as much sanctuary as we can afford.

We can see that there is something of a treadmill effect here. As we solve problems, other problems then take our attention. Some might remark that we can't win, that all problems cannot be solved, so why bother? Yes, we'll always have problems, *but* we're tackling the most severe ones and then moving on to the less severe ones *ad infinitum*. That is progress, and it's making the universe a better place. I want to live in a world where my biggest concern is what to wear today.

MORALITY, KNOWLEDGE, AND TECHNOLOGY

An implicit theme throughout this book is that moral progress and technological progress are intertwined, and that connection is not explicitly acknowledged enough. Obviously, when we have non-animal products superior in every way to animal-based ones, we will have made moral progress, even unwittingly. This is not the only arena where we've seen this result.

During my high school days, I remember the ethical quandary over stem cell research. Stem cells are cells in early development that have not yet specialized for specific roles (e.g., to become a muscle cell). Stem cell research promises new means for regenerating injured tissue and organs. At that point in 2001, all stem cells were harvested from aborted human fetuses. Given that abortion was and is still a hot-button issue, federal funding of this research came under fire, and then-president George W. Bush subsequently restricted it.[512] Fast forward a few years later and scientists figured out a way to have specialized cells convert back into stem cells, circumventing the need for human embryos.[513] Most of the research community now focuses on these induced stem cells, which also offer other advantages such as the fact that we can form the stem cells from the patient in question and ensure better immune system compatibilities.[514] New technology and knowledge allowed us to completely sidestep a moral dilemma.

Similarly, we cannot ignore the facilitative role of technology in the progress of gender equality. In the recent past, devices such as washers, vacuums, and microwaves mean that less time is needed for a homemaker to complete domestic chores.[515] The (presumable) missus was able to find work outside the home and develop a career. We could continue to foster gender equality with technology. An artificial womb would abrogate the biggest biological disadvantage of women when it comes to the modern workplace. Women also tend to bear more of the familial responsibilities taking care of kids and parents,[516] which in turn limits their careers. Technology to delegate such care would also even the playing field of opportunity and flexibility for women. Note: artificial wombs and robotic care strike me as fanciful, intractable technologies. Better policies are a more sensible starting place, as in mandated, long paternity leave, subsidized child care, and societal expectation that men take over more nurturing responsibilities.

In conclusion, we undervalue the generation and spread of knowledge too much; just look at how poorly academic scientists and other teachers are paid. I suspect that this is because the value of innovation is so hard to measure. First, it's hard to associate a piece of knowledge with the ultimate outcome. We can trace individual cases with enough investigation, but a wholesale study that definitively proves that such and such research led to X percent benefit to this industry is difficult to formulate and carry out. It's always going to be a case-by-case basis. Furthermore, the way knowledge seeps and spreads make it difficult to propose a negative control. We'd need to see another part of the Multiverse (Appendix A) where that knowledge wasn't spread. So, it would be difficult to summarize the argument in a scientific paper, which unfortunately remains the prevailing currency for professional acceptance. But that doesn't mean that knowledge isn't one of the most powerful forces for good. We have the explanations for the value of knowledge and that should convince us enough.

CHAPTER TERMS

- **morality**: the solution, reduction, exacerbation, or creation of problems. Better morality is solving and reducing problems; whereas, worse morality is exacerbating and creating problems (such as suffering). We should strive for better morality.
- **relativism:** the untenable idea that morality can be different for different cultures, religions, or species. Given that moral knowledge is like any other knowledge domain (developed through conjectures and refutations), saying morality is relative is akin to saying that each culture can have its own physics or biology.
- **suffering:** prolonged pain in conscious entities. One of the most pressing problems we strive to solve today.

CHAPTER SUMMARY

Morality is about problem-solving versus problem exacerbation/creation. The most moral action is the one that diminishes the total magnitude of all problems. The least is the opposite. We also argue what the problems actually are through the same knowledge generation process (and so what are problems may actually change over time, i.e., we come to learn that animals can experience pain). One of the biggest, hardest-to-refute problems is unnecessary suffering. Therefore, it would be highly moral to diminish animal agriculture as much as possible because of the unnecessary suffering that it thoughtlessly and needlessly entails. We're confident that animals are conscious beings susceptible to pain and suffering. The moral blind spot of animal agriculture has persisted due to how people learn morality (via osmosis from family and peers). That blind spot dissolves upon argumentation and good faith. As we develop new technology, moral issues will become easier to address, and we can then start to tackle problems that seem intractable, if not beyond consideration, such as the suffering of wild animals. Finally, we see that our ability to generate knowledge and technology inexorably affects our ability to solve moral dilemmas because it helps us solve problems. Therefore, generating and spreading knowledge can be a morally-righteous effort to pursue.

CHAPTER TWELVE

Final Thoughts

In my junior year of college, I first decided to give up meat. I had the inchoate sense that meat was bad, and I would do my part by becoming vegetarian. Over the years, I've had my fair share of discussions and well-intentioned arguments that have changed, evolved, and crystallized my thinking. One discussion has stayed with me since my initial foray. My college friend, who I'll call Maynard, asserted to me, "Your choice isn't making any differences in the grand scheme of things. You're just one person among billions." Maynard's point was that everything I was doing was ultimately inconsequential, that the drop of my idealism was diluted and washed away in a vast ocean of reality. At that moment, I was at a complete loss for words. I didn't have an answer then, and these words have remained seared into my memory ever since.

Since then, I've encountered other explanations that push back on Maynard's assertion. For example, while I might not be changing overall production and consumer choices worldwide, I am at least helping individual animals. By *not* consuming meat and later by abstaining from using animal products generally, I have likely spared at least a few animals over my life. That was enough solace for me and continues to motivate me to this day. Still, it seemed like I could be doing more for a cause that needs all the help it can get.

I also learned about S-curve changes in social trends and movements, where there's a slow start up followed by rapidly increasing change until there's a new normal. Earlier, we discussed how marginalized support for same-sex marriage was just a few decades ago. In the span of thirty years, however, that support went from eleven to over fifty percent.[517] Similarly, the percentage of "nones"—people who identify as atheist, agnostic, or non-religious—has increased substantially in the last couple of decades. From 2002 to 2019, the share of "nones" has increased twenty percent for Americans under thirty years old.[518]

Malcolm Gladwell argues in *The Tipping Point* that these S-curves take off due to the motivated players who are well-connected, persuasive, and who employ sticky ideas.[519] We observed this play out on our screens, with influential movies such as *The Birdcage* and *Philadelphia* emphasizing the humanity and plight of homosexuals. The writers, actors, and producers of these films had an outsized impact, helping dispel terrible ideas and instilling better ones, even though collectively these were just a few hundred people at most. Naturally, I see this book as part of Gladwell's point about information and influence spurring the S-curve. And specifically, I hope that the idea of seeing animals as a technology falls into what he deems as "sticky." If all goes well, the ideas here will spread beyond the primary readers of this book and help evolve our conversation about the transition from animal products. I've partly and purposefully included chapter summaries to distill the ideas so that they're easier to reference and pass around and as a basis for discussion.

The other issue with Maynard's assertion is, what's the alternative? If we know that something is a problem and we *can* solve it, then shouldn't we try? At the very least, shouldn't we cease perpetuating the problem? History shows us through many examples that even seemingly insurmountable problems are soluble. We've solved starvation, banished many diseases, and addressed ubiquitous safety issues (think seat belts and airbags). We're continually producing new technology and ideas. The future can continue to get better. But we have to make moves, and we have to acknowledge the problems' gravity.

Problem-solving can also fulfill the "purpose void" left as we have collectively moved away from religion. Only recently have atheism and agnosticism come into wide acceptance.[520] This is a good thing. We're not following the decrees of an egotistical, violent, and fictitious god or gods. We're less beholden to texts that profess inerrancy despite an abundance of internal contradictions.[521] We've adopted progressive values of inclusiveness, free speech, and consideration of all. A collateral effect is the loss of the prescription of that life that we are to follow, a roadmap of how to act toward salvation and eternal happiness in the afterlife. Such a roadmap is reassuring and simple, but it's wrong. Nonetheless, a lack of purpose is a problem.

Instead, I assert that problem-solving *is* this purpose we should follow. Problem-solving can take many forms: taking care of children or older adults; creating media and art to entertain, educate, and enliven others; teaching women in remote villages how to read and self-actualize; entering a lucrative career and donating excess salary to worthy, productive causes such as Effective Altruism. Problem-solving, however, does not always follow a simple flow chart. We have to figure out what the problems are, and we should try to figure out if activities create more problems than they solve, as we've occasionally seen in well-intended non-profit programs.[522] As a result, the problems that we tackle will be different over time, and different people will tackle different areas. All of this to say, there's not a singular life formula for everyone to achieve their purpose. But we can choose to have purpose.

In this book, I've focused on the problem of using animals as a production technology, hopefully convincing you just how awful their use as a food source is, both technologically and ethically. I focus on this topic because I do not believe that it receives attention in proportion to its magnitude. Nonetheless, I'm interested in how we solve problems generally, especially with knowledge generation. Our institutions have much room for improvement in this regard, and I welcome readers to contact me in order to address such efforts. In particular, I'm drawn to the problem of assessing and spreading scientific research as it's clear that the scientific journal system is too outdated and too outmoded for this task, costing us lives and suffering. We can do much better in the age of the internet.

Furthermore, scientists lack adequate understanding of epistemology evidenced by a number of unscientific statements I've lamented (e.g., "There's not enough evidence to make a conjecture or hypothesis here"). Instead, we need a system that incentivizes scientists to make conjectures publicly as well as critiquing existing ones, as I laid out in Chapter 2. I currently envision something like Github,[523] where each problem statement can have conjectures from anyone. If two parties disagree, projects can be forked into separate versions. Presumably, the better version will win out via measurement evidence, falsifiability, and argumentation. Anyway, if a more advanced system to cultivate our knowledge sounds enticing, please reach out to me. If I do write a second book, this will likely be the topic.

Thanks for reading, and I truly appreciate you engaging with and entertaining these ideas. As I said in the introduction, I hope that this is merely the beginning of this conversation; therefore, do not hesitate to reach out, especially if you have disagreements. I would love to hear them.[524] All the best, and I look forward to sharing tapas—topped with better vegan cheese—with you.

Sincerely,

Karthik

Appendix A

Predicting the Future

PREDICTION IN BIOLOGY

"July 29th, 2010, 3am to 4:30am."[525]

It was March 1st, 2010. I was visiting the beachfront University of California in Santa Barbara, interviewing for a graduate student position in Frank Doyle III's lab. Frank's lab was developing mathematical models to predict mass coral reef spawning. In corals, both males and females simultaneously release their gametes, the coral equivalent to sperm and eggs. Female and male gametes find each other by mixing in the ocean currents and fertilize into a coral fledgling called a *planula*. Once the ocean current stills enough, the planula will settle onto a surface and eventually develop into a full coral.

Coral-spawning events are striking and wondrous: a snowstorm imbued with a panoply of colors,[526] and one of nature's true spectacles. Given how precise predictions of coral-spawning events are, one industry offers tourists the opportunity to witness them first-hand. Aspiring admirers can book their flight, cruise, and dive with relative assurance that they will experience the flying, bright planula whipping around them. The timing of the spawning depends on measurable, known quantities: the lunar cycle, the temperature of the water, and ocean salt levels.

Ten years later, I still remember my meeting with Frank. I was wowed at the precision of Frank's model and his unassailable confidence that his lab could predict something as capricious as biological reproduction. Even though I was only a fledgling biologist at that point, I intuited that because biology is so complicated—the congress of more than a million different molecules—that such predictability should be nigh impossible, but Frank's team had accomplished just that. Ultimately, I did not join Frank's lab nor attend the picturesque University of California at Santa Barbara; instead, I attended Northwestern University. However, my reasons were not due to a lack of interest in Frank's work.

After finishing my doctorate at Northwestern University, I joined Uwe Sauer's lab at ETH Zurich in Switzerland in order to commence my postdoctoral research. Uwe continually champions mathematical models in biology: the notion that one of the ultimate thrusts of biological research should be to explain biological phenomena in equations. The benefit of using math would be pronounced, specifically by more precision and engineerability in biology, as in other mature fields. For example, in electrical engineering, a circuit designer is spared from having to know all of the details of the electron flow through the metal and semiconductor material. She can leverage a few equations such as Ohm's law to create a sophisticated device. These equations even form computer-aided design (CAD) software. A computer engineer can design a motherboard, and a chemical engineer can design a production plant all using only CAD software. But we are simply not at that point when it comes to large swaths of design and implementation in biology.

Using math in biological prediction through equations would help accelerate us toward a future without animal products. We could precisely design productive bioprocesses churning out alternative meats and alternative foods (Chapter 3). Also consider if we had equations for the structure of the food and personalized nutrition (Chapters 6). And equations could also reduce the development time toward new, exciting foods (Chapter 8). More ambitiously, could we predict the future without animal products completely? Could we make exact pronouncements of when and which products displace animal-based ones?

Uwe and his lab are pioneers in **metabolomics**, measuring the chemical molecules that make up an organism's metabolism. Immediately before I joined Uwe's lab in Zurich, the lab had just developed a new technology, real-time metabolomics.[527] With real-time metabolomics, one could directly inject live bacteria into a mass spectrometer and blow them into pieces within. The mass spectrometer then separates and resolves the bacteria into its constituent metabolic molecules, including amino acids used to form protein, as well as sugars used for energy and the formation of other cellular components. A single reading on the mass spectrometer would roughly measure how many of these metabolic molecules were inside of the bacteria at a given time. We could cultivate bacteria in vessels of a liquid nutritional medium, and every ten seconds run the mass spectrometer operation. Thereby, we could generate a time profile of how these different molecules changed over time. Uwe figured we could use the technology to study how bacteria "decide" to divide and tasked me accordingly. He theorized that the real-time metabolomics technology would make the problem more tractable. By observing how the counts of these different metabolite molecules changed, we could conceivably find the metabolite molecule within that triggered division once reaching a particular level.

To simplify the research, Uwe and I focused on bacteria that are starved for food and thereby unable to divide. Starved cells were technically conducive to our study because, with a low metabolic background, the metabolic change is more pronounced using our mass spectrometry measurement. We performed real-time metabolomics as these starved bacteria received drops of their preferred nourishment: sugar water. Every drop of sugar elicited a beautiful, pronounced peak in many different metabolic molecules (**Figure 21**). We further observed how the sugar traversed the bacterium's metabolism; in effect, the chemical conversion of sugar into other metabolite molecules, and eventually into larger building blocks (e.g. protein and DNA) constituting the bacterium. Large macromolecules such as protein and DNA are often lumped together and collectively termed the **biomass** of the cell.

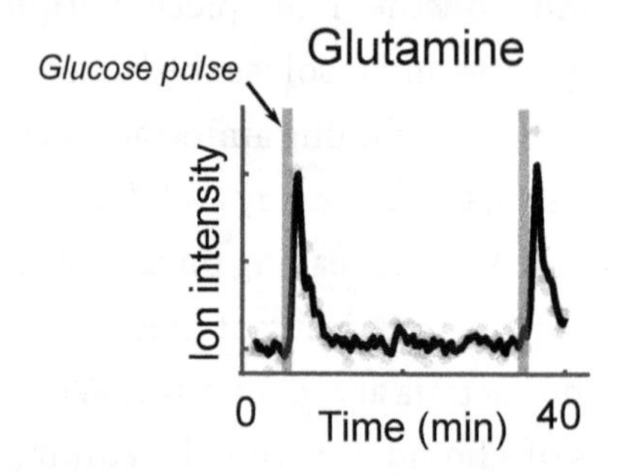

Figure 21. Example of real-time metabolomics data. Drops of sugar were fed to starved bacteria (gray bars). The black line indicates the amount of glutamine (an amino acid used to make protein) in the bacteria's metabolism. The sugar feeding induced a spike in the bacteria's level of glutamine. The consequent fall of the glutamine suggested that the bacteria make protein by rapidly consuming the glutamine.

Uwe and I then hypothesized that the sugar rapidly formed into new biomass, in fact, faster than the ten-second window of our measurement. Follow-up experiments confirmed our suspicions, and the pulsed sugar drops were indeed formed into new biomass of protein and DNA. We then reasoned that division was actually not determined by metabolic level, but instead by biomass level. Eventually, through some parallel experiments, we pinpointed bacterial division to a specific entity within the biomass, a specialized protein, FtsZ. FtsZ comprises the physical ring within the bacterium, affixed to the inner surface. When bacteria divide, this ring constricts, thus squeezing the mother cell into the two daughter cells. Our study highlighted FtsZ as the primary decision point for division; that is, when the cell has enough FtsZ, division commences. With our data and this insight, we were able to develop an equation that predicted when bacteria divided as a function of sugar input. After a scientific review process spanning over a year, the work was finally published.[528]

These two anecdotes might suggest that all events are predictable given the right equation or knowledge, even in a domain with millions of variables such as biology. We simply need to find the right experiment or situation where we can focus on the few variables that matter, and consequently, we can more easily conjecture a model. But as with any explanation, we must always ask if there is an alternative, better explanation. In this case, perhaps coral reef spawning and bacterial division are simply two special cases that are amenable to predictability. Perhaps their predictability actually serves an important biological function.

Predictability as a selection parameter makes sense when you consider another explanation: when the coral release gametes all at once, they maximize their reproductive potential. The simultaneous explosion of gametes floods the environment so that predators are unable to devour everything. If gametes were released piecemeal, then it would be easy picking for predators, and the coral would fail to reproduce. Therefore, predictable coral spawning carries a reproductive advantage and befits evolutionary selection. Presumably, corals whose gamete-release time varied from the synchronized window would readily be selected out of the population as predators would more easily consume their progeny. Therefore, evolution is the forcing function for concerted spawning and explains the predictability here.

For bacterial division, I can't explain as definitively why the quantity of FtsZ drives the division decision. I suspect that FtsZ serves as a proxy indicator for the cell to know how much biomass it has. When the mother cell splits into the two daughter cells, it will all be for naught if the daughters' cells aren't viable. Accordingly, the mother must have sufficient biomass to divvy into two daughters. For example, each daughter cell will need a least one whole copy of the DNA. Each daughter cell will also need some cellular machinery in order to decode and actualize the DNA. Each daughter cell will additionally need requisite amounts of membrane material to encapsulate its contents. Therefore, the mother cell must know when it has enough biomass to grant to each daughter cell. Counting each individual molecule needed seems rather costly, especially given the constraints of bacterial biology. Instead, FtsZ forms more slowly compared to other parts of the cell's biomass, and therefore, it serves as a suitable bellwether for indicating the time to divide. In this scenario, FtsZ is one of the last biomass components to make it to the sufficiency finish line.

STRUGGLING TO PREDICT

In the case of biology, when timing serves an evolutionary function, we should be able to find ways to predict occurrences with enough investigation. But the ultimate question still looms: are there limits to what else we can predict generally? Could I predict which products will replace animal ones and invest my money accordingly? When I started my scientific career, I thought that everything would be predictable given more knowledge. I believed that we could eventually simulate everything with perfect fidelity, only limited by the performance of our computers. Indeed, some events are readily predictable after we perform the key measurements of the determinants. Astronomical events with celestial objects occur like clockwork. Hotel bookings were sold years before The Great American Eclipse of 2017.[529] We know that Halley's comet will appear in our sky every seventy-four to seventy-nine years.

However, the case of weather is more vexing when it comes to our confidence in our prediction ability. We actually have the equations we can use already. Weather is highly defined by the movement of air and water, calculable using the Navier-Stokes equations. The Navier-Stokes equations are scripturally elegant:

$$\rho \frac{D\vec{V}}{Dt} = -\nabla p + \rho \vec{g} + \mu \nabla^2 \vec{V}$$

These equations predict how fluid flows at large scales. Being able to solve for these equations would conceivably detail the path and progression of fluid flow. However, the equations are notoriously difficult to use, and solving them has proved insuperable except for simple cases. In 2000, The Clay Mathematics Institute offered a million-dollar prize for proving just the *existence* of unique, smooth analytical solutions to these equations.[530] So far, there have been no awardees. An idealized, analytical Navier-Stokes solution would take the form of a function *u*(*x*, *y*, *z*, *t*) = *<solution>*, where u is the speed of the flow at a given position, defined by the spatial x, y, and z coordinates at a certain time t. With a solution, we would be able to calculate the speed of a fluid at any time and position.

Even with the lack of analytical solutions, the Navier-Stokes equations are not completely inaccessible to us. We can use them numerically; so, say, start with a value (e.g., the speed of a fluid) and calculate the predicted speed after a short time. We can perform this process iteratively: after we get the speed at the next time, we can plug it in again to get the speed at the time point after that. Seemingly, we could perform calculation *ad infinitum*. However, numerical solutions always bear some error because we're using numbers with a finite precision, i.e., ceasing the numbers after the decimal place. The error only amplifies with each further iteration. We intuit that if our error after five iterations is ten percent, then after five more iterations, the values will have error of more than ten percent, perhaps even twenty percent. Indeed, the error can only get worse: as we continue to forecast our calculations, we eventuate to an unbounded solution, where the error becomes so large that the calculated values become meaningless. Numerical solutions always carry this disadvantage, though they are still functionally useful within those limits.

The tendency to blow up—reach unboundedness—is compounded with the Navier-Stokes equations because they are **chaotic**. Chaotic systems and equations are extremely sensitive to initial conditions. I highlighted that the ideal solution would take the form $u(x, y, z, t)$ = <solution>. Expounding further, a solution would depend on the initial conditions, i.e., how fast the fluid is moving at the start. For instance, in the case of fluid flow, consider the situation where a gust of wind travels into a valley. In the middle of the valley, a tall rock blocks the path, and so the wind flows around it. A simple schematic is shown in **Figure 22A**, where the valley is the light gray area, the wind is the arrows, and the rock is the white circle in the middle. The wind requires about five minutes to reach full speed. In the first minute, the wind is imperceptible, but at around 1.5 minutes, the wind picks up, and by three minutes in, the wind is nearly at full speed. Visually, I depict such changing wind with a graph (**Figure 22B**). The wind speed rapidly ramps between the 1.5- and 3-minute period.

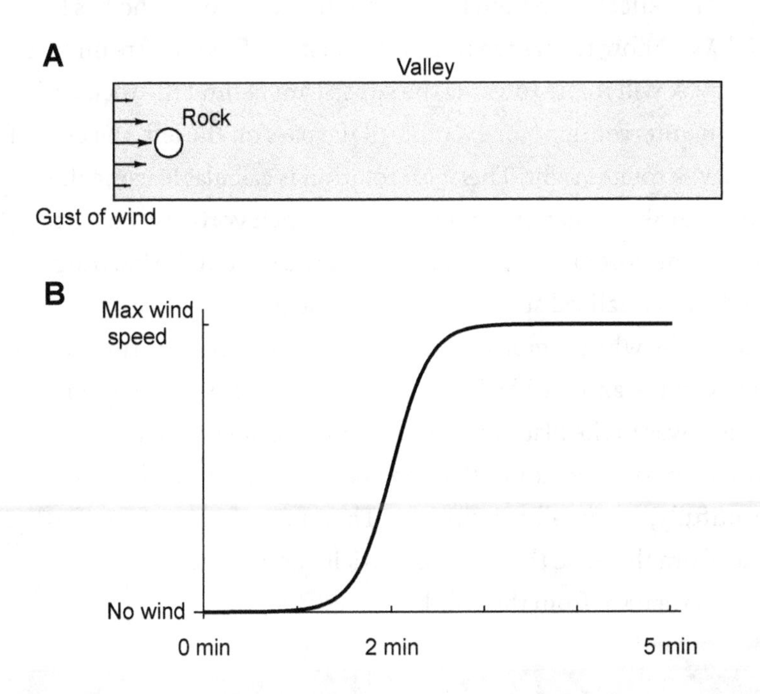

Figure 22. An example for predicting fluid flow. (A) A depiction of a valley with a rock (indicated by the white circle). Wind flows from the west side heading eastward. (**B**) The profile of the wind speed over five minutes. Initially the wind is modest, but quickly ramps to maximum speed between 1.5 to 3 minutes.

The wind cannot flow through the rock; it must flow around it. As a result, circulating eddies are formed where anything traveling along with the wind will rotate. Imagine that we're riding a leaf carried by the same wind. Our leaf skirts past the right side of the rock just barely. The wind is diverted from impact with the rock and thereby creates a rotation at specific points in the valley. Our leaf now spins leftward and as we move past the rock, we now look back at it before being rotated back looking into the valley. This local rotational effect from fluid flow is familiar to anyone who has kayaked. We know that as we ride past a rock in a flowing stream that our kayak will rotate towards the slipstream behind the rock without any intervention; for example, if we pass on the left, the bow of the kayak rotates right. This local rotation is calculable using the Navier-Stokes equations and formally termed **vorticity**. I have calculated the vorticity at 4.5 minutes into the start of the blowing wind and visualized such in **Figure 23**. Indeed, we can see the eddies. The white regions indicate high local rotation turning counterclockwise, and the black regions indicate clockwise rotation. If our leaf was in the black region, the wind would be rotating us clockwise. We also notice that the vorticity zigzags further out beautifully, eventually dissipating. Therefore, if our leaf was further away from the rock, there is less vorticity and, therefore, less applied rotation from the wind.

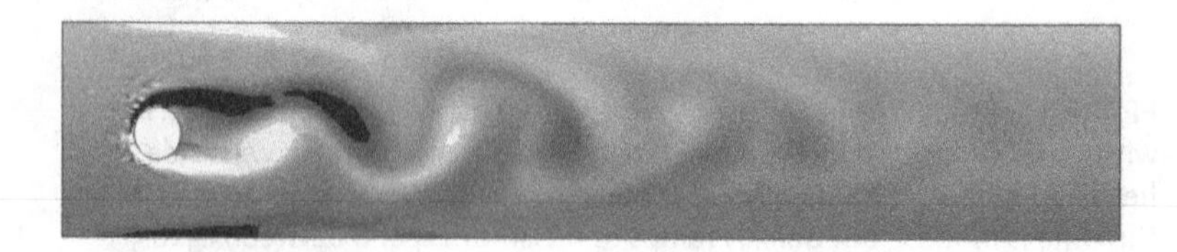

Figure 23. The vorticity of the wind moving past the rock. The intensity of vorticity is indicated by the coloring where black indicates turning clockwise and the white indicates turning counterclockwise.

Vorticity drastically affects our weather, as the local spinning can help lift (advect) moist air into our atmosphere.[531] If advection seems unintuitive, consider a tornado, which picks up material along its path. The dirt is kicked up and moves upward as the tornado spins. With weather, when enough moist air is concentrated, the clouds will precipitate by releasing water in the form of rain, snow, sleet, or hail. Therefore, we are not able to predict whether it will rain without accounting for vorticity. However, as I alluded to earlier, the calculations are sensitive to how the situation starts due to the chaotic properties of the Navier-Stokes equations. To illustrate, consider if we slightly changed the ramping of the wind ever so slightly in **Figure 24A**. I am not changing the speed, just adjusting the time when the wind starts to pick up by half a second (New Case 1) or almost a full second (New Case 2). In fact, the change is so imperceptible that we need to zoom in significantly to perceive the difference as the wind speeds ramp up. However, at three minutes in, all three cases are indistinguishable, as observable by the same zoom level at that point.

So how does this affect the consequent vorticity? When looking at the same time point of 4.5 minutes, the vorticities indeed look quite different (**Figure 24B**). We see that New Case 1 and 2 look to be much further along compared to the original Base Case. The black region in the Base Case seems to be just budding off, whereas, in the new cases, this budded region is further along. Just to emphasize, at 4.5 minutes, the blowing wind from all three cases should be indistinguishable because it was so at three minutes. Furthermore, we only shifted the start of the ramp by about a half second for each case. Fluid parameters such as vorticity are indeed highly sensitive to the starting conditions.

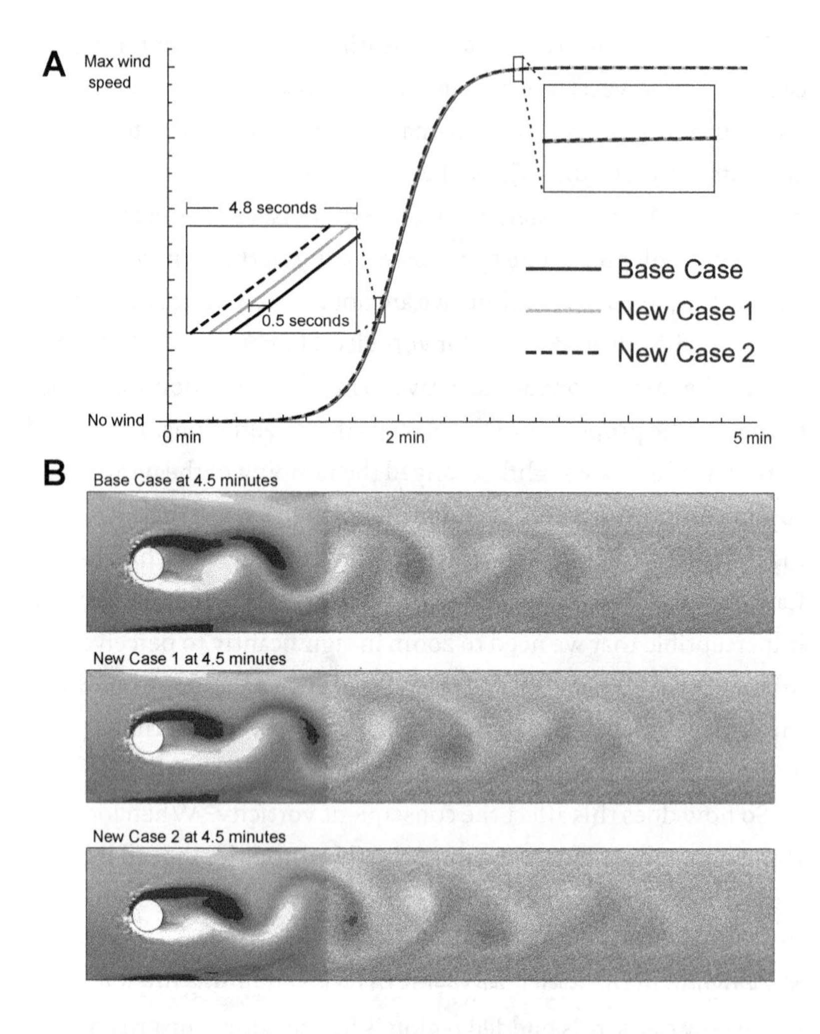

Figure 24. Consequences of changing the ramp start time. (**A**) Two additional cases with earlier ramp times were added (New Case 1=gray line, New Case 2=dashed line). The change in ramp time is slight—about 0.5 seconds for each new case, as depicted within the zoom in. By three minutes, the profiles are seemingly identical, even with the same zoom as the first. (**B**) Change in vorticity for the new cases. The slight change in ramp time elicits a new vorticity profile at 4.5 minutes.

Notice how most weather predictions never proceed beyond a couple of weeks? That's because numerical solutions of the Navier-Stokes equations underlie such predictions, and as we've just seen, they can be difficult to precisely use. Certainly, vorticity can also be affected by the rotation of the Earth or seasonal effects; therefore, a measly wind may be inconsequential if it is the dry season in India. But consider a season where there is some possibility for rain, such as in Durham, North Carolina during the spring. Effects from gusts of wind may prove the deciding factor, and our calculations are precarious enough that rain is forecast in probabilities: "A thirty percent chance of rain."

The chaotic nature of the Navier-Stokes even limits a potential analytical solution, too. Suppose that a special mathematician comes along, solves the equations, and claims his or her million dollars from the Clay Institute. With the solved equations, we would still require a perfectly precise measurement of the starting conditions (like the ramping of the wind). Errors in this measurement will only propagate the further that we extend our solution. We saw how divergent the solutions became after just a couple of minutes. Consider hours, days, or weeks later. Clearly, we cannot assure a clear day for, say, that outdoor wedding six months out that we want to plan.

Even if the mathematics of the Navier-Stokes equations was perfectly manageable, it's still an approximation to the underlying phenomenon. We know that water is comprised of individual molecules that coalesce into fluid. The equations do not account for this molecular resolution. There is an inherent limit to where they work. This is okay for predictions of weather because knowing all the details of the individual molecules is unnecessary for making predictions. We can **abstract** away those features, and we do not have to account for the random motions of particulate in the water nor the vapor cavities collapsing and inducing shockwaves in the proximal region of the flow. While these phenomena undoubtedly influence the fluid flow at microscopic levels, we ignore them for larger-scale predictions.

Should we desire to study the physical world more granularly, we require new models to explain these scales. All known matter is comprised of molecules; molecules themselves are arrangements of atoms. Zooming in further, atoms are defined by elementary particles including electrons and neutrons. Photons, another elementary particle, form light, similar to how water molecules form bulk fluid flow. Elementary particles can exhibit different, discrete states, for example having a defined location or speed. Curiously, the elementary particle seems more like a cloud or a wave rather than a single point before we try to measure a property such as speed or position. There is an ensemble of different property values. Mathematically, this cloud is termed a **wave function**. Upon "measurement," this wave function collapses, and a single point remains with a defined position value. The measurement can be repeated, drawing different values, but is bracketed by the wave function. This variance in different values has been famously termed the Heisenberg Uncertainty Principle.

MULTIVERSE THEORY

This peculiarity of quantum physics has perplexed scientists for years, and a range of explanations has ensued. The instrumentalist interpretation sees the wave function collapse as purely mathematical and divorced from physical reality. The pervasive, timid Copenhagen interpretation shirks away from complete physical explanation in that it does not try to say what the wave function physically is. According to the Copenhagen interpretation, the photon only has a position or speed *after* measurement; before that, one should throw one's hands up in the air. Both lines of thinking are highly problematic. If we're to be consistent with our use of scientific knowledge, we have to look for a physical interpretation of before and after the wave function collapse. We cannot excuse for inconvenience, and we must boldly seek to find such an interpretation.

In 1957, physics graduate student Hugh Everett resolved the interpretation issue of the wave function collapse and proposed the many-worlds interpretation (**Multiverse theory**) of quantum physics. According to Everett's theory, expounded on later by physicist David Deutsch, the cloudy elementary particle is a real physical ensemble of each electron, proton, photon, etc., in the different states enumerated by the wave function, i.e., having different positions and speeds according to the mathematical distribution. The wave function is *physically* real. Within the ensemble, each different state is fundamentally **fungible** with each other. It is like having two one-dollar bills in your wallet, where one is smooth and another is crumpled. If you're buying a pack of gum, you may reach into your wallet and either of the bills will satisfy the payment. Therefore, in the context of the payment, both bills are fungible with one another. Similarly, the particles, while having different states, are fungible to the eventual measurement. Once the measurement occurs, one of the fungible particles within the ensemble is chosen and represents the collapse of the wave function.

Measurement here means that the particle is **entangled** with the current reality, meaning that other particles are affected by the state of the particle. In the money analogy, once the bill has been handed over and placed into the cash register it affects whatever is around it. The many-world interpretation takes this a step further, positing that innumerable realities are actually branching from one another. Each fungible state of the particle is entangled with a different reality. To return to our money analogy, it's as if the money transfer is the source point for two timelines: one where the gum was paid for with the smooth bill, and another paid for with the crumpled. Both timelines occur and are indeed real but are inaccessible to one another. In the same way, every time a particle is entangled into a reality, the complement reality occurs as well. There is another reality where the other state of the particle is chosen and entangled to create an altogether separate timeline.

Seemingly the best explanation for quantum wave phenomena, the Multiverse theory's implications are astounding. The branching occurs so often that it's effectively infinite. Our lives are no longer defined by a single narrative. There may be timelines where we marry completely different people, die young or old, don't exist at all, etc. The collective aggregate of all of these timelines and histories is termed the Multiverse.

To reify the potential consequence of such quantum variance and branching, take for example light, which is comprised of the elementary particle, photons. The photons emitted from the sun directly land and penetrate to some degree the epithelial cells of our skin. So let's imagine Stella, a hypothetical gal, sunbathing on the beach where she is bombarded with radiating light photons all day. In one part of the Multiverse, a photon, in the corresponding state of position and speed, is able to induce a lesion in the DNA of Stella's skin.[532] This lesion in Stella's DNA facilitates the genesis and progression of skin cancer. In the other part (and likely by far the majority of causalities) of the Multiverse, Stella sunbathes without incident. Consider ramifications of Stella getting the cancer or not. Suppose she was on the path to develop the killer application of the alternative food movement. Her getting cancer set back the movement a year or two and cedes market placement to another researcher and company.

The Stella example is imaginably exceptional, as it's one photon inducing significant, propagating changes in the affected timelines; however, it effectively illustrates the potential for continuous divergence. Single elementary particles can also induce electronics failures,[533] but for the most part, the elementary particles only create variances in aggregate. So, more than a million subatomic particles make up a single FtsZ protein, and it takes two thousand to divide a cell. Therefore, it's hard to say exactly what phenomena are subject to the variance wrought by subatomic particles collapsing into different states. I suspect something as large as Halley's comet is mostly buffered against quantum effects. The momentum of impinging light or particulate is unlikely to steer the comet off track substantially, no matter which part of the Multiverse we're in. We'll be able to predict the entry of Halley's comet in our night's sky with high confidence.

Going back to our case of predicting weather; as discussed, the Navier-Stokes equations are chaotic and subject to wildly different outcomes with even the slightest changes in the starting point, as highlighted in the vorticity example earlier. David Deutsch asserts in *The Fabric of Reality* that this flow of fluid is subject to these variances resulting from the Multiverse branching.[534] There is no such thing as a perfect measurement of wind ramp-speed due to quantum fluctuations. Furthermore, the quantum variance within fluid flow amplifies due to the chaotic effects. More pointedly, there are defined limits to weather predictability: weather will always vary in different parts of the Multiverse due to the marginal starting differences. We *cannot physically* predict weather completely.

So, it seems that there are clearly phenomena that we can predict (e.g., arrival of Halley's comet) and phenomena that we can't (e.g., whether rain occurs on the same day next year, how and when the alternative food revolution takes root). Predictability is graded and lies on a spectrum. There are limits to the fidelity of prediction.

Particularly, I'm fascinated with our minds in this regard. Our thought processes are the outgrowth of a network of elongated cells, neurons. Neurons pass information to each other through electrical current in the form of action potentials. When neurons are excited enough—when an electrical voltage induces a strong enough change, the action potential occurs, and downstream neurons can also be activated. Action potentials are "all or nothing." The neurons are designed in a way that the voltage must cross a specific threshold in order for the action potential to occur. If the voltage remains below the threshold, the activation will not occur. Seemingly, this mechanism could be susceptible to quantum variance, and I'm not the first to suggest this.[535] Therefore, in different parts of the Multiverse, our thinking could be different for the same situation. The words I type out now are surely different compared to the words I type in another part of the Multiverse. That is clearly consequential and should steer how our world unfolds.

KNOWLEDGE AND PROSPERITY

We can't discuss the science of predictability without the most obvious application: the stock market. The promise of financial windfall entices smart, intrepid individuals to predict the future of companies, industries, and commodities. Hedge funds and managed funds pool investment from a variety of sources: institutions, rich investors, and retail investors. Such funds are generally run by a team who will charge the participating investors a flat one to three percent of the total investment every year. This charge compensates the team for their expert prognosticating efforts derived from all their research activity and predictive models. Alternatively, one can invest in index funds: an index fund collectively invests in companies by some set parameter. For example, investing in the SPY index fund pegs the money to the performance of the Standard & Poor's 500, the 500 largest companies in the United States. No expert customizes the fund allocation. The SPY is investing directly into the biggest 500 companies, that's it. One can even just directly assert that the entire stock market will succeed and invest in index funds such as VTSAX—a total stock market fund offered by Vanguard, a consumer-focused investment firm. Index funds do not charge the exorbitant fees of a managed fund because investors are not paying anyone for anything other than administration, so typically the cost is a mere fraction of a managed fund.

So, which is better for an average retail investor like me? Index funds by a wide margin: in the last twenty years, only around five to eight percent of actively managed funds beat the SPY.[536] So, it would seem that trying to predict winners in the market is more akin to trying to predict the weather long-term than the orbit of Halley's comet. There are probably many Stella situations where consequential situations unfold differently in various parts of the Multiverse, leading to different economic outcomes. And I suspect that human thinking is susceptible to the branching from all these collapsing waves. It seems best for retail investors to avoid the managed funds, save on the high fees, and place their money in index funds.

Curiously, we're *still* better off investing the money in the stock market, whether index or managed, than putting it in a savings account or other "safer" strategies. The S&P 500 itself has averaged about ten percent growth per year since 1926.[537] Major hedge funds average at around eight percent growth[538] in recent years, and the US economy averages around five to six percent. Certainly, these numbers are not adjusted for inflation—our money devalues by about two to three percent per year—but nonetheless, our money is better invested than sitting under our mattress. This growth must be coming from somewhere. The companies are not always vampires sucking the wealth from people. In order for that to be true, everyone else's lives would have to be getting much worse over time, and as discussed in the first chapter, that just isn't true, globally speaking. The assertion that wealth is merely extracted from the planet is also untenable, as the US has been consuming fewer total resources with a growing population, in certain sectors.[539] Instead, this growth ultimately comes from the net creation of wealth, particularly in the form of new knowledge.

I'm not the first to make such a point. Nassim Nicholas Taleb makes a similar assertion in his book *Black Swan*: market economies have more inherent opportunities for serendipities—wondrous new findings that spur cascade effects.[540] Yuval Harari Noah in *Sapiens* asserts that the growth of our economies can be best explained by the advancement of scientific knowledge.[541] When we have scientific techniques that reduce the overall magnitude of problems, we grow the market. For example, the impactful Haber process allows us to derive fertilizer from the air. The process, developed in the early 20th century, solved the biggest agriculture problem at the time, supplying enough nitrogen to the plants. In fact, the Haber process has been so consequential that an estimated 2 billion people or so may not have existed without the technology.[542] And again, the Haber process is actually making fertilizer out of air. The cost is not exactly zero—one needs a catalyst, reactor, and the proper temperature/pressure, but clearly, we can derive much benefit for meager input. The Haber process then has ripple effects. Fewer people need to be farming, and we can accordingly divert resources to other problems.

We notice a similar effect with the eradication of smallpox. The poverty of a population is reduced by no longer having to put resources into fighting this debilitating disease. Altogether, as we solve problems using knowledge, our prosperity as a modern civilization should increase.

A large swath of modern, prosperity-inducing inventions have come from the industrialized West (e.g., the Americas and Europe), including innovations in health, agriculture, and computing. Admittedly, the West has more resources to put toward such problems; but, how did the West get these ample resources in the first place? In this, we have to recognize a valuable difference between liberal, free-market societies versus authoritarian ones (e.g., Soviet-style communism, old monarchies and tyrannies). Author Yuval Noah Harari also discusses the difference in both *Sapiens* and *Homo Deus*, analogizing the economy in terms of computers, that is, a market-based economy has more "computers" running calculations. These computers are the different companies in one industry, and their calculation is each one's specific business model. Having a variety of different business models to try means that the best ones will rise, and the subpar ones will die out. In contrast, an authoritarian-run economy runs only one computer or a few, as determined by the state. This computer is stuck running one business model for the given industry, and that business model is not stress-tested against others. As a result, a better business model will not be found because of the lack of recourse and competition.

But is that the only explanation? There's a country with ostensible communist leanings that seems to buck this trend: China. From 1979 to 2010, China's entire economy has averaged ten percent growth per year; it's like the entire country is the S&P 500.[543] So why does the Chinese economy seemingly perform counter to our intuitions? Certainly, China has liberalized over the years by, for example, allowing private companies to supplant the state-owned enterprises, but I think there are additional explanations. Our classic sense of a lumbering, inefficient socialist government brings to mind the Soviet Union. China's government and economy are structured differently. The Soviet Union directed and micromanaged from the top whereas China's government is more confederated.[544] Municipal governments have wide leeway to try things differently compared to other municipalities. China, in effect, runs a lot of computer calculations, but the calculations occur locally within the individual municipalities.

Secondly, China invests heavily in infrastructure. Sticking with the computer analogy, the infrastructure is the computer hardware that enables more calculations. Roads and bridges facilitate the transport of goods, thus permitting the industry within to burgeon. Public transit enables easier commuting and thus creates more opportunities for workers. Schools and hospitals engender a healthy, educated populace who can pursue opportunities more easily and spur more innovations. Infrastructure as a stimulus only works to a point. After a certain number of roads and hospitals, there is diminishing marginal value in adding more. Beyond that point, other forces must take root to facilitate further growth. It'll be interesting to see how China navigates the post-infrastructure investment binge.

Finally, China can reverse decisions and policies easily even compared to a dynamic society such as the US. Consider the 2020 coronavirus pandemic. China initially and irresponsibly played down notions of a brewing epidemic to its own citizens and the world at large.[545] Eventually, China's central government wholly acknowledged the situation and imposed restrictions on the movement of citizens, helping contain the spread of the virus.[546] And it worked. In contrast, the American federated system allowed each state to pursue its own action toward the pandemic, each with varying responses of shelter-in-place, closing schools, vaccination, and reopening businesses. The result became a disaster as the United

States had higher rates of coronavirus cases and deaths than most of the world.[547] Stamping out a pandemic requires strong coordination which is much easier with top-down control.[548] Solving policy issues such as pandemics does not suit an American model of governance.

Understanding and refuting bad policies promotes progress. Normally, in a liberal democracy, as in the West, a referendum on policy comes through the ouster of politicians via an election, the bedrock of democracy. Such political party turnover doesn't happen in China, and for most authoritarian governments, this would normally be a huge problem. The lack of punishment for political failures has ensured tepid progress in politically repressive governments such as Russia, Zimbabwe, and Iran, whose leaders cannot be ousted so easily and, therefore, have less reason to falsify their actions, even though they have the power to do so readily. In contrast, Chinese government officials are held accountable, just not via elections. In rural areas, officials are held accountable through membership in their local temple and lineage groups.[549] In order to maintain high member standing, the official must perform activities for the community such as building roads and schools. Moreover, urban officials will be dismissed, sometimes jailed, upon the revelation of corruption or aberrant behavior.

This accountability ultimately results in the willingness to acknowledge and revoke bad policies. Officials have phased out the One Child Policy, which famously only allowed one child per family. China has also taken strides to address some of the corruption issues raised during the Tiananmen Square protests; it developed some of the most comprehensive anti-corruption laws to date.[550] And most recently, China is leading worldwide diplomatic efforts for CO_2 emission reduction after years of heavy industrial pollution.[551] All in all, having more consolidated control can be an advantage for ridding bad policies or proceeding in the best direction. *Being able to move on from clearly inferior ideas creates innovation.*

But there is a caveat. When we repudiate an idea or policy, a new one (including no policy) takes its place. This process allows people to propose potentially superior ideas. Here, the West has a much stronger advantage given the expansive, impressive free speech protections, where citizens are not punished for critiquing governments, businesses, or the actions of others. Therefore, I sincerely chide China for its strict, limiting speech laws that likely muffle the creativity of their populace. Western-style free speech protection ensures that the best ideas can be generated and that inferior ones can be weeded out. Furthermore, amazement at China's ability to innovate does not excuse their poor human rights record. I severely condemn the Uighur re-education camps and jailing of protestors. These are immoral and create more problems ultimately.

In sum, I agree that scientific advancement and serendipitous findings underlie the growth of our economies and the total reduction of problems. For example, replacing animal technology would solve many problems in terms of nutrition, environment, suffering, taste, and cost as we discuss throughout the book. However, we can go one step further. I assert that scientific advancement equates to the repudiation of bad ideas while generating new ones. Therefore, if we employ more of this process—knowledge generation—within our society, our economy should grow.

Circling back to index funds: index funds allow one to invest in entire markets. It doesn't matter which company or academic institution generates the next Haber process. The ripple effects create a new positive, and the overall market efficiency grows, thereby creating new wealth and increasing the value of investments. Therefore, as long as humanity generates knowledge, I see the economic output growing and the value of investments rising. This is the best reason that I see to invest in index funds. The knowledge that confers efficiencies throughout the marketplace can be reaped via index funds without picking the specific winners (the next Apple or Tesla). Furthermore, we also do not have to be in the right part of the Multiverse to realize a profit. I assert that all parts of the Multiverse from the current branch point will continue to generate knowledge.

I will happily continue to invest in markets where I suspect that untrammeled knowledge generation can occur. I do not have too much hope for countries such as Iran, which suppresses free speech and its female population's ability to add to its innovation capacity. In general, theocracies are unable to innovate as readily as more open societies.[552] By definition, religious doctrine is inviolate and does not allow itself to be falsified no matter how bad the idea, whether it's mutilating genitalia or condoning slavery. Furthermore, theocracies and religion do not reward the generation of new ideas to replace old ones, but rather demand complete obedience to established ones.

PREDICTION OF TECHNOLOGICAL TRENDS

With a more precise idea of how innovation works and how our societies pursue breakthroughs, we can now make a key prediction: innovations will happen under the right circumstances (e.g., free speech, academic research, repudiation of bad ideas) and will be also driven by needs (e.g.. better transportation capability). The causalities of innovation will always be the less efficient and dragging technologies. For example, the development of modern thermodynamics by academic scientists led to combustion engines.[553] As a result, motorized vehicles were developed, and the use of horse-drawn carriages waned. We see similar innovation and needs driving superannuated technology in other areas: going to a video rental store is a nuisance when we can sit on our couch and rent with clicks on our remote, thanks to streaming capacities of the internet. Light-emitting diode bulbs are more energy-efficient and cooler than old school incandescent. And online services (e.g., filing taxes) tend to overtake in-person or snail-mail–based ways.

Less saliently, we might not think about cows as a technology, but we certainly use them as such. They are used as a reactor that consumes grass and generates meat and dairy. If we take this idea further, the notion of growing a bag of meat that walks around and takes four years to mature sounds pathetic. We could develop hardy, local bioreactors that perpetually supply enough "meat" to feed a village. We could have 3D printers in each of our homes that are programmed to our specification and "print" out a meat patty on demand, perfectly catered to our taste and nutrition at that moment. We could have a bag of protein powder that, when poured into water, self-assembles into a steak. Even before any of those futures, in the grocery stores we should have non-animal meat in every way superior to animal protein: cheaper, more nutritious, and better tasting. This should be possible because animals are such crummy technology.

Unfortunately, we have not innovated enough yet, but we will. And I posit that *knowledge generation is the key to rendering animal technology obsolete*. With the right knowledge, we will have no need for animals as a technology. Chapters 2 through 5 are about making that case and showing that animals are ripe for replacement in terms of the biological physics, our ability to improve on them, and the constraints of food processes. Therefore, provided that our capacity for innovation continues, we will render animal products obsolete. Innovation is replacing poorer technology with better technology. The future becomes a present without animal products.

I will not predict when. In fact, I defer to the Multiverse theory as the best knowledge for explaining the wave function collapse. Therefore, I take its claims seriously in that our universe is constantly branching, and there are many instances of myself typing this same sentence. Therefore, if I try to predict when, then I may appear brilliant in one part of the Multiverse and silly in another.

I also cannot predict how. The nature of knowledge generation is inherently unpredictable. Predicting future knowledge is tautological, for, if one could perfectly predict new knowledge, then we would have that knowledge. Given that I theorize that we'll innovate ourselves away from using animal products, the replacement may look completely different across the branches of the Multiverse. In one part, the personalized 3D printers may dominate, and in another, drones may be flying in our steak from a local bioreactor. I can only say that if we continue to innovate and produce new knowledge, animals will eventually be replaced because they are nakedly terrible technology.

CHAPTER TERMS

- **metabolomics:** measuring large swaths of an organism's metabolism
- **biomass:** the physical mass of a biological organism
- **chaotic:** a property of a system where future states are highly sensitive to the starting conditions of the initial values of the variables
- **vorticity:** the localized, imparted net rotation of a fluid. Imagine a flower petal in a stream of water and notice how it turns at various points of the stream without colliding with any objects. This rotation stems from the vorticity of the fluid.

- **abstraction**: purposefully discounting details that are inconsequential or unnecessary when explaining a phenomenon
- **fungible**: interchangeability of a group of objects and entities. Most common example is money. If I pay taxes but don't want it to fund the military, I'm out of luck. The taxes go into a common pool, from which the military draws its funding.
- **wave function**: a fungible state of subatomic particles exhibiting different states (in particular, speed and position). Upon measurement (entanglement), one of the particles is chosen.
- **Multiverse**: the total ensemble of all of the different timelines created by the wave function collapse. This can mean realities where Hillary Clinton was elected in 2016, where Soviet Officer Vasili Arkhipov was not able to avert nuclear war in 1962, or where Genghis Khan fell off his horse and died before expanding the Mongol empire.
- **entangle**: after a wave function collapse, the particle in question interacts with other particles, setting in consequences

CHAPTER SUMMARY

We can predict some events with appreciable fidelity (e.g., the orbits of Halley's comet), but other events are fundamentally unpredictable (e.g., the weather a year from today or which companies do well over the next twenty years) due to living within the Multiverse, constraints on physical laws (e.g., fluid flow), and the nature of knowledge generation. The Multiverse theory, currently the best explanation for quantum effects, suggests that all events are constantly branching through advancing time. There are realities where we do not exist, and the idea of a defined narrative is fallacious. Despite this inability to precisely pinpoint the future, certain trends are predictable from the generation of knowledge, including that poorer knowledge and technology will be replaced. This trend continually creates wealth and means that we can reap rewards from just investing into knowledge-generating economies. Animals are a susceptible technology and long due for replacement.

Acknowledgments

This book is dedicated to my parents. They have cultivated within me with the independence, dedication, and confidence that I needed to write this book. Likewise, they've encouraged me to follow my own decisions, never mandating what I should do. They've been wholly supportive in just about every decision I've made in my life, even as they're gritting their teeth as I spend weekends writing versus wooing a potential wife. I would be nowhere without them, and I will be eternally grateful. I also thank my sister, Kavya, who is unafraid to intellectually spar with me and helped develop my thinking on these topics. She was the piece of flint that kindled *After Meat* as showcased in the introductory story in Chapter 1.

I would like to thank my postdoctoral research mentor, Uwe Sauer. Never before has anyone been so honest and direct with me. I was jarred and disheartened at first, but ultimately, I became a better scientist and thinker under his mentorship. Appreciating the positive impact, the values of honesty and directness are thoroughly instilled throughout this book. In Uwe's lab, I befriended Elad

Noor. Elad is one of the deepest, most abled thinkers I've ever met, and many ideas in *After Meat* can be pinpointed to a conversation with him. Similarly, I thank my doctoral advisor, Keith EJ Tyo. Despite nominally being an "applied" scientist, Keith continually champions scientific understanding and implores students not merely to be satisfied with what they've been told—they need to probe deeper and find additional explanations.

I thank my editors Brandon Crowther, Michael Sanders, and Genevieve Morgan. Brandon reviewed and edited the initial manuscript after every three chapters were written. He positively affected the trajectory of the book and helped clarify, refine, and even repudiate many of the arguments. I thank Michael Sanders for helping take the beta-version of this book to a polished, well-produced work. Michael had the overwhelming task of bringing forth the big picture of a complicated tome that spanned many disparate, difficult topics. Only then could he help with the reshaping and specific revisions that ensued afterward. Not many editors can do what Michael so deftly accomplished, and I'm sincerely thankful to be connected to him. Genevieve consumed a lengthy, complicated work in a short amount of time. She sharpened the writing, improved the explanations, and reinforced the arguments.

I thank a number of beta readers who provided feedback on a preliminary version of *After Meat* including Jeffray Behr, Ilana Taub, Elad Noor, Sarah Edwards, Nate Crosser, Sarah May, Peter Su, Susan J. Miller, Edward Candales, Casey Riordan, Blake Byrne, Will Roderick, Devi Sekar, Bryanna Foote, L. Vinod Reddy, Lauren Adeelar, and Anonymous Reader One.

Many have also helped with the researching of this book. Karol Orzechowski of Faunanalytics provided valuable resources about the manner in which animal carcasses are divvied up and distributed. Ron Sender and Ron Milo provided a lot of literature and help with the nitrogen balance section. I thank Amy Huang of the Good Food Institute for helping me understand the state of funding for alternative foods at the academic level.

I also thank many individuals who helped to push me to write and publish this book. I was incredibly fortunate to connect with Blake Byrne of the Good Food Institute. Blake has connected me to helpful contacts, and our routine conversations have seeped into the pages here. I was fortunate enough to meet vegan fashion designer Jill Fraser in my local writers' group. She has taken whatever lengths she can to connect me with many to help on this endeavor. Mark Warner, Paul Shapiro, Jacy Reese, Magnus Vinding, Gidon Eshel, Sascha Camilli, Anna Akbari, and Beth Clevenger have discussed the publication process with me in some way, directly or indirectly affecting the end product. I thank Cary Hayner, Jesse Lou, Sacha Laurin, Alifia Merchant, and Kavya Sekar for helping me with publicity preparation. Frank Doyle III and Suckjoon Jun were kind of enough to grant the use of their names in Appendix A and Chapter 9, respectively.

Lastly, I would not have been emboldened to pursue this book without some intellectuals and non-fiction authors pioneering the way. Peter Singer energized the animal welfare movement with his 1975 book *Animal Liberation*. Even though I only read the book after becoming vegan, it's obvious that Singer's ideas have diffused their way into me and my thinking. His efforts demonstrate how knowledge can be absorbed even without direct contact. Sam Harris has convinced me that one can indeed win over people with superior arguments and non-fiction books. Harris also taught me that knowledge isn't siloed in specific domains; we should all freely explore other areas, even outside of our ostensible purview. Yuval Noah Harari wrote two impressive books (*Sapiens* and *Homo Deus*) that draw insightful inferences from looking at many domains in tandem, a modus influential to *After Meat*. Harari has also highlighted the problem of animal agriculture prominently. Steven Pinker has buoyed my feelings about human progress, that we have indeed made improvements and continue to do so. Pinker is also the best non-fiction writer I've encountered. He's inspired me to write clearly and to respect the intelligence of my audience. Finally, David Deutsch's *The Beginning of Infinity* has upended my worldview entirely; it changed how I think about how new knowledge/technology is generated, how we humans are uniquely positioned to solve problems and drive progress, and how I want to lead my life from now on. I still think about *The Beginning of Infinity* nearly every day. In many ways, *After Meat* can be viewed as a progeny of *The Beginning of Infinity*, taking direct lessons to tackle the problems of animal technology.

About the Author

KARTHIK SEKAR, PHD

Karthik was born and raised in North Carolina, USA. He has a bachelor's degree in biomedical engineering from the University of North Carolina, a doctorate in chemical engineering from Northwestern University, and postdoctoral research experience from ETH Zurich. Karthik's research career has spanned many topics related to the future of food, including bioreactor design, bioengineering, metabolism, and systems/quantitative biology. In his spare time, Karthik enjoys hiking, trying new recipes, playing board games, listening to podcasts, meditating, and reading non-fiction.

List of Figures and Tables

CHAPTER 6

CHAPTER 7

CHAPTER 8

APPENDIX A

Index

Symbols

A

B

C

D

E

H

I

J

K

L

M

N

O

P

Q

R

S

T

U

Endnotes and References

INTRODUCTION

1 Carman, T. (2019). Burger King Is Rolling out a Meatless Whopper. Can McDonald's Be Far behind? Washington Post. **https://www.washingtonpost.com/news/voraciously/wp/2019/04/01/burger-king-is-rolling-out-a-meatless-whopper-can-mcdonalds-be-far-behind/** (Accessed November 1, 2020).

Heil, E. (2019). McDonald's Gets into the Veggie Burger Game, Testing a Beyond Meat Patty in Canada. Washington Post. **https://www.washingtonpost.com/news/voraciously/wp/2019/09/26/mcdonalds-gets-into-the-veggie-burger-game-testing-a-beyond-meat-patty-in-canada/** (Accessed November 1, 2020).

2 Parker, J. (2019). The Year of the Vegan. Economist. **https://www.bluehorizon.com/economist-names-2019-the-year-of-vegan/** (Accessed July 12, 2020).

3 I can be reached at http://aftermeatbook.com/contact.

CHAPTER 1

4 Humira Price. True Med Cost. **https://www.truemedcost.com/humira-price/** (Accessed December 14, 2019).

5 Adalimumab - Drug Usage Statistics, ClinCalc DrugStats Database. ClinCalc.com. **https://clincalc.com/DrugStats/Drugs/Adalimumab** (Accessed November 1, 2020).

DailyMed - IMRALDI- Adalimumab Injection, Solution. U.S. National Library of Medicine. **https://dailymed.nlm.nih.gov/dailymed/drugInfo.cfm?setid=acdfaa71-27ed-4717-8e7d-f1a5fe0d1fa6** (Accessed November 1, 2020).

6 Quianzon, C. C., & Cheikh, I. (2012). History of insulin. Journal of Community Hospital Internal Medicine Perspectives, 2(2), 18701. **https://doi.org/10.3402/jchimp.v2i2.18701.**

7 Holland, J. W. (2008). Chapter 4—Post-translational modifications of caseins. In A. Thompson, M. Boland, & H. Singh (Eds.), Milk Proteins (pp. 107–132). Academic Press. **https://doi.org/10.1016/B978-0-12-374039-7.00004-0.**

8 McEvoy, M. (2019). Organic 101: What the USDA Organic Label Means. U.S. Department of Agriculture. **https://www.usda.gov/media/blog/2012/03/22/organic-101-what-usda-organic-label-means** (Accessed November 3, 2020).

9 Healthy Lives | Natural Diet - How to Eat More Naturally. Healthy Lives. **https://www.healthylives.com/kb/natural-diet/** (Accessed January 18, 2021).

10 Rogers, O. (2019). The 4 Ns Of Meat-Eating. Faunalytics. **https://faunalytics.org/the-4-ns-of-meat-eating/** (Accessed November 3, 2020).

11 USDA Organic. U.S. Department of Agriculture. **https://www.usda.gov/topics/organic** (Accessed January 18, 2021).

12 Mayo Clinic Staff. (2019). Paleo diet: What is it and why is it so popular? Mayo Clinic. **https://www.mayoclinic.org/healthy-lifestyle/nutrition-and-healthy-eating/in-depth/paleo-diet/art-20111182** (Accessed December 26, 2019).

13 Lieberman, D. E., Venkadesan, M., Werbel, W. A., Daoud, A. I., Dandrea, S., Davis, I. S., Mangeni, R. O., & Pitsiladis, Y. (2010). Foot strike patterns and collision forces in habitually barefoot versus shod runners. Nature, 463(7280), 531–535. **https://doi.org/10.1038/nature08723**.

14 Kirk Cameron Tells Piers Morgan Homosexuality Is 'Unnatural,' 'Ultimately Destructive.' (2016). HuffPost. **https://www.huffpost.com/entry/kirk-cameron-piers-morgan-homosexuality-unnatural_n_1318430** (Accessed December 26, 2019).

15 Tomasello, M., Melis, A. P., Tennie, C., Wyman, E., & Herrmann, E. (2012). Two key steps in the evolution of human cooperation: The interdependence Hypothesis. Current Anthropology, 53(6), 673–692. **https://doi.org/10.1086/668207**.

16 Fertility Rate, Total (Births per Woman) | Data. The World Bank. **https://data.worldbank.org/indicator/SP.DYN.TFRT.IN** (Accessed December 26, 2019).

17 Fertility Rate, Total (Births per Woman) - Bangladesh. The World Bank. **https://data.worldbank.org/indicator/SP.DYN.TFRT.IN?locations=BD** (Accessed December 26, 2019).

18 Harari, Y. N. (2018). Sapiens: A Brief History of Humankind. Reprint edition. New York: Harper Perennial.

19 CDC. (2019). Heart Disease Facts. Centers for Disease Control and Prevention. **https://www.cdc.gov/heartdisease/facts.htm** (Accessed December 26, 2019).

20 Sachs, J. (2006). The End of Poverty. Economic Possibilities for Our Time. Reprint edition. New York, NY: Penguin Books.

21 This metric is normalized (adjusted) for inflation and purchasing power. Specifically, it's defined as $1.90 a day (2011 PPP). Poverty Headcount Ratio at $1.90 a Day (2011 PPP) (% of Population) - World | Data. The World Bank. **https://data.worldbank.org/indicator/SI.POV.DDAY?locations=1W&start=1981&end=2015&view=chart** (Accessed December 26, 2019).

22 Poverty Headcount Ratio at $1.90 a Day (2011 PPP) (% of Population) - China| Data. The World Bank. **https://data.worldbank.org/indicator/SI.POV.DDAY?locations=CN** (Accessed December 26, 2019).

23 Elwood, J. M. (1989). Smallpox and its eradication. Journal of Epidemiology & Community Health, 43(1), 92–92. **https://doi.org/10.1136/jech.43.1.92.**

24 WHO | Smallpox. (2007). WHO Media centre. **https://web.archive.org/web/20070921235036/http://www.who.int/mediacentre/factsheets/smallpox/en/** (Accessed December 26, 2019).

25 Cool Earth | Working to Support Rainforest Communities to Reduce Deforestation. **https://www.coolearth.org/** (Accessed December 15, 2019).

26 MacAskill, W. (2016). Doing Good Better: How Effective Altruism Can Help You Help Others, Do Work That Matters, and Make Smarter Choices about Giving Back. Reprint edition. Avery.

CHAPTER 2

27 Rodríguez-Bravo, B., Nicholas, D., Herman, E., Boukacem-Zeghmouri, C., Watkinson, A., Xu, J., … Świgoń, M. (2017). Peer review: The experience and views of early career researchers. Learned Publishing, 30(4), 269–277. **https://doi.org/10.1002/leap.1111.**

28 Popper, K. (2002). Conjectures and Refutations: The Growth of Scientific Knowledge. 2nd edition. London ; New York: Routledge.

29 Gibbs, J. W. (1961). The Scientific Papers of J.Willard Gibbs, Vol. 1: Thermodynamics. Dover Publications.

30 Yehuda, R., Daskalakis, N. P., Bierer, L. M., Bader, H. N., Klengel, T., Holsboer, F., & Binder, E. B. (2016). Holocaust Exposure Induced Intergenerational Effects on FKBP5 Methylation. Biological Psychiatry, 80(5), 372–380. **https://doi.org/10.1016/j.biopsych.2015.08.005**.

31 ArXiv.Org e-Print Archive. **https://arxiv.org/** (Accessed November 28, 2020).

32 ChemRxiv: The Preprint Server for Chemistry - Browse. **https://chemrxiv.org/** (Accessed November 28, 2020).

33 MedRxiv.Org - the Preprint Server for Health Sciences. **https://www.medrxiv.org/** (Accessed November 28, 2020).

34 BioRxiv.Org - the Preprint Server for Biology. **https://www.biorxiv.org/** (Accessed November 28, 2020).

35 Flier, J. S. (2020). Covid-19 Is Reshaping the World of Bioscience Publishing. STAT. **https://www.statnews.com/2020/03/23/bioscience-publishing-reshaped-covid-19/** (Accessed November 28, 2020).

36 Kodvanj, I., Homolak, J., Virag, D., Trkulja, V., & Trkulja, V. (2020). Publishing of COVID-19 Preprints in Peer-reviewed Journals, Preprinting Trends, Public Discussion and Quality Issues. BioRxiv. **https://doi.org/https://doi.org/10.1101/2020.11.23.394577**.

37 Deutsch, D. (2012). The Beginning of Infinity: Explanations That Transform the World. Reprint edition. New York NY Toronto London: Penguin Books.

38 Carreyrou, J. (2018). Bad Blood: Secrets and Lies in a Silicon Valley Startup. 1 edition. New York: Knopf.

39 Smil, V. (2005). Creating the Twentieth Century: Technical Innovations of 1867-1914 and Their Lasting Impact. 1 edition. Oxford; New York: Oxford University Press.

40 Dennett, D. C. (1992). Consciousness Explained. 1 edition. Boston: Back Bay Books.

41 Deutsch, David. (2012). The Beginning of Infinity: Explanations That Transform the World. Reprint edition. New York NY Toronto London: Penguin Books.

42 Ke, Q., Ferrara, E., Radicchi, F., & Flammini, A. (2015). Defining and identifying Sleeping Beauties in science. Proceedings of the National Academy of Sciences of the United States of America, 112(24), 7426–7431. **https://doi.org/10.1073/pnas.1424329112.**

43 Chakma, J., Sun, G. H., Steinberg, J. D., Sammut, S. M., & Jagsi, R. (2014). Asia's Ascent — Global Trends in Biomedical R&D Expenditures. New England Journal of Medicine, 370(1), 3–6. **https://doi.org/10.1056/NEJMp1311068.**

44 Projected global automotive research and development expenses between 2012 and 2020. Statista. **https://www.statista.com/statistics/373853/global-automotive-research-and-development-spending/** (Accessed December 26, 2019).

45 Darby, Kenyatta, and Matthew World War II Rubber Crisis/ Invention of Synthetic Rubber. The History Of Rubber. **http://historyofrubber.weebly.com/wwii.html** (Accessed December 26, 2019).

46 Estimates of Funding for Various Research, Condition, and Disease Categories (RCDC). **https://report.nih.gov/categorical_spending.aspx** (Accessed December 26, 2019).

47 IFPMA. (2017). THE PHARMACEUTICAL INDUSTRY AND GLOBAL HEALTH FACTS AND FIGURES 2017. International Federation of Pharmacutical Manufacturers & Associations, 86. **https://www.ifpma.org/wp-content/uploads/2017/02/IFPMA-Facts-And-Figures-2017.pdf.**

48 Holden. (2014). Returns to Life Sciences Funding. The GiveWell Blog. **https://blog.givewell.org/2014/01/15/returns-to-life-sciences-funding/** (Accessed December 26, 2019).

CHAPTER 3

49 Product Design. (2020). Wikipedia. **https://en.wikipedia.org/w/index.php?title=Product_design&oldid=978871905** (Accessed November 8, 2020).

50 Panzarino, M. (2012). This Is How Apple's Top Secret Product Development Process Works. The Next Web. **https://thenextweb.com/apple/2012/01/24/this-is-how-apples-top-secret-product-development-process-works/** (Accessed December 22, 2019).

Murphy, M. (2018). What It Was like Working at Apple to Create the First iPhone. Quartz. **https://qz.com/1380188/ken-kocienda-qa/** (December 22, 2019).

51 What Is the Design Process? James Dyson Foundation. **https://www.jamesdysonfoundation.com/content/dam/pdf/Standalone_DesignProcess.pdf.**

52 Nicolas Leblanc | French Chemist. Encyclopedia Britannica. **https://www.britannica.com/biography/Nicolas-Leblanc** (Accessed December 22, 2019).

53 Katz, D. L. (1980). History of Chemical Engineering, Advances in Chemistry, Series 190.

54 Manufacturing of Sodium Carbonate by Solvay Process. (2014). worldofchemicals.com. **https://www.worldofchemicals.com/440/chemistry-articles/manufacturing-of-sodium-carbonate-by-solvay-process.html** (Accessed November 12, 2020).

55 Chocolate Crystals, and the Chemistry behind Tempering Chocolates. Chocolate Tempering Machines. **https://www.chocolatetemperingmachines.com/pages/chemistry-behind-tempering-chocolates** (Accessed January 19, 2021).

56 Hayward, J. (2000). Too Little, Too Late: An Analysis of Hitler's Failure in August 1942 to Damage Soviet Oil Production. The Journal of Military History, 64(3), 769. **https://doi.org/10.2307/120868.**

Operation Barbarossa. (2019). Wikipedia. **https://en.wikipedia.org/w/index.php?title=Operation_Barbarossa&oldid=931734815** (Accessed December 22, 2019).

57 Shapiro, P., and Harari, Y.N. (2018). Clean Meat: How Growing Meat Without Animals Will Revolutionize Dinner and the World. 1st edition. New York: Gallery Books.

58 McCollough, J., & Check, H. F. (2010). The baleen whales' saving grace: The introduction of petroleum based products in the market and its impact on the whaling industry. Sustainability, 2(10), 3142–3157. **https://doi.org/10.3390/su2103142.**

59 Sedentism. (2019). Wikipedia. **https://en.wikipedia.org/w/index.php?title=Sedentism&oldid=921051968** (Accessed December 22, 2019).

60 Russell, N. (2002). The wild side of animal domestication. Society and Animals, 10(3), 285–302. **https://doi.org/10.1163/156853002320770083.**

61 Alvard, M. S., & Kuznar, L. (2001). Deferred Harvests: The Transition from Hunting to Animal Husbandry. American Anthropologist, 103(2), 295–311. **https://doi.org/10.1525/aa.2001.103.2.295.**

62 Vann, M. (April 10, 2009). A History of Pigs in America. **https://www.austinchronicle.com/food/2009-04-10/764573/** (Accessed December 22, 2019).

63 Scanes, C. G. (2018). The Neolithic Revolution, Animal Domestication, and Early Forms of Animal Agriculture. Elsevier Inc. **https://doi.org/10.1016/B978-0-12-805247-1.00006-X.**

64 Burrow, H. M., Moore, S. S., Johnston, D. J., Barendse, W., & Bindon, B. M. (2001). Quantitative and molecular genetic influences on properties of beef: A review. Australian Journal of Experimental Agriculture, 41(7), 893–919. **https://doi.org/10.1071/EA00015.**

65 Burrow, H. M., Moore, S. S., Johnston, D. J., Barendse, W., & Bindon, B. M. (2001). Quantitative and molecular genetic influences on properties of beef: A review. Australian Journal of Experimental Agriculture, 41(7), 893–919. **https://doi.org/10.1071/EA00015.**

66 Speth, J. D. (2010). Big-Game Hunting in Human Evolution: The Traditional View. In J. D. Speth (Ed.), The Paleoanthropology and Archaeology of Big-Game Hunting: Protein, Fat, or Politics? (pp. 39–44). Springer. **https://doi.org/10.1007/978-1-4419-6733-6_3.**

67 Allen, R., & Allen, R. C. (2007). How Prosperous were the Romans? Evidence from Diocletian`s Price Edict. Economics Series Working Papers, (363).

68 Orchard, J. E., & Buck, J. L. (1938). Land Utilization in China. Geographical Review, 28(4), 698. **https://doi.org/10.2307/210316.**

69 "List of Countries by Meat Consumption." (2019). Wikipedia. **https://en.wikipedia.org/w/index.php?title=List_of_countries_by_meat_consumption&oldid=930341394** (Accessed December 22, 2019).

Coming from mainly the following reference:

Current Worldwide Annual Meat Consumption per capita, Livestock and Fish Primary Equivalent. Food and Agriculture Organization of the United Nations.

70 Smil, V. (2013). Should We Eat Meat?: Evolution and Consequences of Modern Carnivory (1st edition). Wiley-Blackwell.

71 MacDonald, J. M., & McBride, W. D. (2011). The Transformation of U.S. Livestock Agriculture Scale, Efficiency, and Risks. SSRN Electronic Journal, (43). **https://doi.org/10.2139/ssrn.1354028.**

72 Transparency Market Research. (2017). Heparin Market to Reach a Valuation of US$ 16.3 Billion by the End of 2025: Transparency Market Research. GlobeNewswire News Room. **http://www.globenewswire.com/news-release/2017/11/08/1177610/0/en/Heparin-Market-to-Reach-a-Valuation-of-US-16-3-Billion-by-the-End-of-2025-Transparency-Market-Research.html** (Accessed December 22, 2019).

73 Gladwell, M. (2017). "Revisionist History Season 2 Episode 9." Revisionist History. **http://revisionisthistory.com/episodes/19-mcdonalds-broke-my-heart** (Accessed December 22, 2019).

74 Marti, D. L., Johnson, R. J., & Mathews, K. H. (2012). Where's the (Not) meat? Byproducts from beef and pork production. There's The Beef: Select Research on Global Beef Production and Trade, 55–81.

75 Tyson Foods, Inc. (2018). Form 10-K. United States Securities and Exchange Commission
Specifically, page 25.

76 Apple Operating Margin 2006-2019 | AAPL. (2019). MacroTrends. **https://www.macrotrends.net/stocks/charts/AAPL/apple/operating-margin** (Accessed December 22, 2019).

77 Coca-Cola Profit Margin 2006-2019 | KO. (2019). MacroTrends. **https://www.macrotrends.net/stocks/charts/KO/coca-cola/profit-margins** (Accessed December 22, 2019).

78 The History Behind Bovine Serum Albumin. (2017). Lifecycle Biotechnologies. **https://www.lifecyclebio.com/history-behind-bovine-serum-albumin/** (Accessed December 22, 2019).

79 Impossible Foods. Impossible Foods. **https://impossiblefoods.com/** (Accessed December 22, 2019).

80 National Beer Sales & Production Data. Brewers Association. **https://www.brewersassociation.org/statistics-and-data/national-beer-stats/** (Accessed December 22, 2019).

81 History of Beer. Heartland Brewery. **https://www.heartlandbrewery.com/history-of-beer/** (Accessed December 22, 2019).

82 Historie. Weihenstephaner Brewery. **https://www.weihenstephaner.de/die-brauerei/historie/** (Accessed December 22, 2019).

83 Orth, J. D., Thiele, I., & Palsson, B. O. (2010). What is flux balance analysis? Nature Biotechnology, 28(3), 245–248. **https://doi.org/10.1038/nbt.1614**.

84 I calculated using the values as follows:
Start with the pitch rate from the following source:
Pitch Rates. Wyeast Laboratories. **https://wyeastlab.com/pitch-rates** (Accessed December 22, 2019).
So (100 to 300 billion yeast pitched into a beer fermentation).
Each yeast is roughly 20 picogram per this source:
Size and Composition of Yeast Cells - Budding Yeast Saccharomyces Ce - BNID 108315. BioNumbers. **https://bionumbers.hms.harvard.edu/bionumber.aspx?id=108315&ver=3&trm=yeast+mass&org=** (Accessed December 22, 2019).

85 Lei, H., Feng, L., Peng, F., & Xu, H. (2019). Amino Acid Supplementations Enhance the Stress Resistance and Fermentation Performance of Lager Yeast During High Gravity Fermentation. Applied Biochemistry and Biotechnology, 187(2), 540–555. **https://doi.org/10.1007/s12010-018-2840-1**.

86 My calculations are detailed and available here: **https://github.com/karsekar/fwoap/blob/master/book/exponentiallinear/notebook.ipynb**.

87 Doubling Time of 'Normal' Laboratory Haploid - Budding Yeast Saccharomyces Ce - BNID 108255. BioNumbers. **https://bionumbers.hms.harvard.edu/bionumber.aspx?id=108255&ver=5&trm=yeast+synthetic+complete+medium&org=** (Accessed December 22, 2019).

88 Oduah, E. I., Linhardt, R. J., & Sharfstein, S. T. (2016). Heparin: Past, present, and future. Pharmaceuticals, 9(3), 1–12. **https://doi.org/10.3390/ph9030038.**

89 Wiebe, M. (2004). Quorn™ Myco-protein — Overview of a successful fungal product. Mycologist, 18(1), 17–20. **https://doi.org/10.1017/S0269-915X(04)00108-9.**

90 Moore, D., Robson, G. D., & Trinci, A. P. J. (2020). 21st Century Guidebook to Fungi (2nd edition). Cambridge University Press. **http://www.davidmoore.org.uk/21st_Century_Guidebook_to_Fungi_PLATINUM/Ch17_18.htm** (Accessed December 2, 2020).

91 Wiebe, M. (2002). Myco-protein from fusarium venenatum: A well-established product for human consumption. Applied Microbiology and Biotechnology, 58(4), 421–427. **https://doi.org/10.1007/s00253-002-0931-x.**

92 Ritala, A., Häkkinen, S. T., Toivari, M., & Wiebe, M. G. (2017). Single cell protein-state-of-the-art, industrial landscape and patents 2001-2016. Frontiers in Microbiology, 8(OCT). **https://doi.org/10.3389/fmicb.2017.02009.**

CHAPTER 4

93 Appl, M. (2011). Ammonia, 2. Production Processes. In Ullmann's Encyclopedia of Industrial Chemistry. American Cancer Society. **https://doi.org/10.1002/14356007.o02_011.**

94 Note that in order to properly compare yield of gasoline-based cars to electric vehicles, we have to calculate from the same origin. First let's start with fossil fuel. Gasoline cars take the fossil fuel directly (yielding at generally 20% energy efficiency). Electric vehicles are powered by the grid, which is often powered by fossil fuels. Even though the grid is relatively efficient in reaping energy from fossil fuels, there is additional loss when that's passed to an electric vehicle. There is only about 25% energy yield for electrical cars, in aggregate, if the grid is fossil fuel powered. However, over 50% yield is achievable for electrical cars if the grid is provided by solar or wind sources. If we try to calculate the yield for gasoline cars in the same way, it'll be in the single digits or less percentage. Remember fossil fuels are pressured animal carcasses, so they ultimately come from solar sources too, just far, far less efficiently. Sources:

- All-Electric Vehicles. United States Department of Energy. **http://www.fueleconomy.gov/feg/evtech.shtml** (Accessed December 5, 2020).
- Metcalfe, M. (2018, July 7). Grid Efficiency: An Opporunity to Reduce Emissions. Energy Central. **https:/energycentral.com/c/ec/grid-efficiency-opportunity-reduce-emissions** (Accessed December 5, 2020).

95 Keefe, T. J. (2012, April 23). The Nature of Light. **https://web.archive.org/web/20120423123823/http:/www.ccri.edu/physics/keefe/light.htm** (Accessed April 11, 2021).

96 Insect - Circulatory System. Encyclopedia Britannica. **https://www.britannica.com/animal/insect** (December 15, 2019).

97 My calculations, data, and analyses are written up here and are freely available: **https://github.com/karsekar/fwoap/blob/master/book/maturationtime/notebook.ipynb.**

98 I used the Stokes-Einstein equation to calculate the diffusion constant (assume water at 25 degrees Celsius). We'll assume that dioxygen is a 0.3 nanometer radius, so that's a diffusion constant approximately equal to $1.5*10^{-9}$ meters^2/s. Now, we can calculate the time to diffuse using the diffusion time equation (for 2 micrometers, or roughly the length of *E. coli* bacterium) and obtain about 2-3 milliseconds. Sources for these values:

Einstein Relation (Kinetic Theory). (2019). Wikipedia. **https://en.wikipedia.org/w/index.php?title=Einstein_relation_(kinetic_theory)&oldid=929563873** (Accessed December 11, 2019).

Milo, R., & Phillips, R. (2015). Cell Biology by the Numbers (1st edition). Garland Science. **http://book.bionumbers.org/how-big-is-an-e-coli-cell-and-what-is-its-mass/** (Accessed December 11, 2019).

99 West, G. B., Brown, J. H., & Enquist, B. J. (2001). A general model for ontogenetic growth. Nature, 413(6856), 628–631. **https://doi.org/10.1038/35098076.**

100 West, G. (2017). Scale: The Universal Laws of Growth, Innovation, Sustainability, and the Pace of Life in Organisms, Cities, Economies, and Companies (First Edition). Penguin Press.

101 My calculations, data, and analyses are written up here and are freely available: **https://github.com/karsekar/fwoap/blob/master/book/yield/notebook.ipynb.**

102 Van Iersel, M. W. (2003). Carbon use efficiency depends on growth respiration, maintenance respiration, and relative growth rate. A case study with lettuce. Plant, Cell and Environment, 26(9), 1441–1449. **https://doi.org/10.1046/j.0016-8025.2003.01067.x.**

103 Liu, Y., Yang, Y., Wang, Q., Du, X., Li, J., Gang, C., ... Wang, Z. (2019). Evaluating the responses of net primary productivity and carbon use efficiency of global grassland to climate variability along an aridity gradient. Science of the Total Environment, 652, 671–682. **https://doi.org/10.1016/j.scitotenv.2018.10.295.**

104 Lötscher, M., Klumpp, K., & Schnyder, H. (2004). Growth and maintenance respiration for individual plants in hierarchically structured canopies of Medicago sativa and Helianthus annuus: The contribution of current and old assimilates. New Phytologist, 164(2), 305–316. **https://doi.org/10.1111/j.1469-8137.2004.01170.x.**

105 Shu, S. miao, Zhu, W. ze, Wang, W. zhi, Jia, M., Zhang, Y. yuan, & Sheng, Z. liang. (2019). Effects of tree size heterogeneity on carbon sink in old forests. Forest Ecology and Management, 432(June 2018), 637–648. **https://doi.org/10.1016/j.foreco.2018.09.023.**

106 Kempes, C. P., Dutkiewicz, S., & Follows, M. J. (2012). Growth, metabolic partitioning, and the size of microorganisms. Proceedings of the National Academy of Sciences of the United States of America, 109(2), 495–500. **https://doi.org/10.1073/pnas.1115585109.**

107 Why Is Saffron so Expensive? Quora. **https://www.quora.com/Why-is-saffron-so-expensive** (Accessed December 15, 2019).

108 Johnson, W. (2001) Prelude to the Irish Famine: The Potato. Ireland Story. **https://www.wesleyjohnston.com/users/ireland/past/famine/potato.html** (Accessed December 15, 2019).

109 Irish Potato Famine. (2019, February 19). Extra Credits. **https://www.youtube.com/watch?v=gAnT21xGdSk** (Accessed December 15, 2019).

110 Woodham-Smith, C. (1992). The Great Hunger: Ireland: 1845-1849 (Reissue edition). Penguin Group.

111 Smil, V. (2013). Should We Eat Meat?: Evolution and Consequences of Modern Carnivory (1st edition). Wiley-Blackwell.

112 White, M. C. (2016, March 20). Why Whole Foods Is Saying No to Fast-Growing, Beefed-Up Chickens. Money. **https://money.com/whole-foods-chicken-breeding/** (Accessed December 15, 2019).

113 Thanksgiving turkeys can't have sex because their breasts are too big. (2011, November 17). Grist. **https://grist.org/article/2011-11-17-thanksgiving-turkeys-cant-have-sex-because-their-breasts-are-too/** (Accessed December 15, 2019).

114 Lin, Linly. (2015, November 27). Fattest-Ever U.S. Cattle Herd Signals End to Record Beef Prices. Bloomberg. **https://www.bloomberg.com/news/articles/2015-11-27/fattest-ever-u-s-cattle-herd-signals-end-to-record-beef-prices** (Accessed August 3, 2021).

115 Foer, J. S. (2010). Eating Animals. Back Bay Books.

116 My calculations are here: **https://github.com/karsekar/fwoap/blob/master/book/productivity/notebook.ipynb**.

117 Kleiber, M. (1947). BODY SIZE AND METABOLIC RATE. Physiological Reviews, 27(4), 511–541. **https://doi.org/10.1152/physrev.1947.27.4.511**.

118 Celebrated Charities That We Don't Recommend. (2009). The GiveWell Blog. **https://blog.givewell.org/2009/12/28/celebrated-charities-that-we-dont-recommend/** (Accessed December 18, 2019).

119 The water benefit would primarily come from savings. A bioreactor will provide more food for less water than something like a goat.

120 Savir, Y., Noor, E., Milo, R., & Tlusty, T. (2010). Cross-species analysis traces adaptation of Rubisco toward optimality in a low-dimensional landscape. Proceedings of the National Academy of Sciences of the United States of America, 107(8), 3475–3480. **https://doi.org/10.1073/pnas.0911663107**.

121 Thornton, P. K., & Herrero, M. (2010). Potential for reduced methane and carbon dioxide emissions from livestock and pasture management in the tropics. Proceedings of the National Academy of Sciences of the United States of America, 107(46), 19667–19672. **https://doi.org/10.1073/pnas.0912890107**.

122 Thornton, P. K., & Herrero, M. (2010). Potential for reduced methane and carbon dioxide emissions from livestock and pasture management in the tropics. Proceedings of the National Academy of Sciences of the United States of America, 107(46), 19667–19672. **https://doi.org/10.1073/pnas.0912890107.**

123 Pegurier, E. (2017) Study Links Most Amazon Deforestation to 128 Slaughterhouses." Mongabay Environmental News. **https://news.mongabay.com/2017/07/study-links-most-amazon-deforestation-to-128-slaughterhouses/** (Accessed January 18, 2020).

124 Shoval, O., Sheftel, H., Shinar, G., Hart, Y., Ramote, O., Mayo, A., … Alon, U. (2012). Evolutionary Trade-Offs, Pareto Optimality, and the Geometry of Phenotype Space. Science, 336(6085), 1157–1160. **https://doi.org/10.1126/science.1217405.**

125 Shoval, O., Sheftel, H., Shinar, G., Hart, Y., Ramote, O., Mayo, A., … Alon, U. (2012). Evolutionary Trade-Offs, Pareto Optimality, and the Geometry of Phenotype Space. Science, 336(6085), 1157–1160. **https://doi.org/10.1126/science.1217405.**

126 Towbin, B. D., Korem, Y., Bren, A., Doron, S., Sorek, R., & Alon, U. (2017). Optimality and sub-optimality in a bacterial growth law. Nature Communications, 8. **https://doi.org/10.1038/ncomms14123.**

127 You, C., Okano, H., Hui, S., Zhang, Z., Kim, M., Gunderson, C. W., … Hwa, T. (2013). Coordination of bacterial proteome with metabolism by cyclic AMP signalling. Nature, 500(7462), 301–306. **https://doi.org/10.1038/nature12446.**

128 Schuetz, R., Zamboni, N., Zampieri, M., Heinemann, M., & Sauer, U. (2012). Multidimensional optimality of microbial metabolism. Science, 336(6081), 601–604. **https://doi.org/10.1126/science.1216882.**

129 Towbin, B. D., Korem, Y., Bren, A., Doron, S., Sorek, R., & Alon, U. (2017). Optimality and sub-optimality in a bacterial growth law. Nature Communications, 8. **https://doi.org/10.1038/ncomms14123.**

130 Specht, L., & Crosser, N. (2020). State of the Industry Report - Fermentation: An Introduction to a Pillar of the Alternative Protein Industry. Good Food Institute. **https://gfi.org.**

131 "How Is Fungi-Based Meat Made?." Meati Foods. **https://meati.com/pages/how-is-meati-made** (Accessed January 19, 2021).

132 Wiebe, M. (2004). Quorn™ Myco-protein — Overview of a successful fungal product. Mycologist, 18(1), 17–20. **https://doi.org/10.1017/S0269-915X(04)00108-9.**

133 Moore, D., Robson, G. D., & Trinci, A. P. J. (2020). 21st Century Guidebook to Fungi (2nd edition). Cambridge University Press. **http://www.davidmoore.org.uk/21st_Century_Guidebook_to_Fungi_PLATINUM/Ch17_18.htm** (Accessed December 2, 2020).

CHAPTER 5

134 Myth Buster: Edison's 10,000 attempts. (2012). Edisonian, 9. **http://edison.rutgers.edu/newsletter9.html#4** (Accessed December 7, 2019).

135 Chu, J. (2007). A Safe and Simple Arsenic Detector. MIT Technology Review. **https://www.technologyreview.com/s/407222/a-safe-and-simple-arsenic-detector/** (Accessed December 26, 2019).

136 Steinbüchel, A., & Füchtenbusch, B. (1998). Bacterial and other biological systems for polyester production. Trends in Biotechnology, 16(10), 419–427. **https://doi.org/10.1016/S0167-7799(98)01194-9.**

137 How Did They Make Insulin from Recombinant DNA? National Institute of Health. **https://www.nlm.nih.gov/exhibition/fromdnatobeer/exhibition-interactive/recombinant-DNA/recombinant-dna-technology-alternative.html** (Accessed December 26, 2019).

138 Uhlen, M., & Ponten, F. (2005). Antibody-based proteomics for human tissue profiling. Molecular and Cellular Proteomics, 4(4), 384–393. **https://doi.org/10.1074/mcp.R500009-MCP200.**

139 Moran, L. (2016). How Many Proteins in the Human Proteome? Sandwalk. **https://sandwalk.blogspot.com/2016/12/how-many-proteins-in-human-proteome.html** (December 7, 2019).

140 Zahn-Zabal, M., Michel, P.-A., Gateau, A., Nikitin, F., Schaeffer, M., Audot, E., ... Lane, L. (2019). The neXtProt knowledgebase in 2020: data, tools and usability improvements. Nucleic Acids Research, (April 2011), 1–7. **https://doi.org/10.1093/nar/gkz995**.

141 Mahieu, N. G., & Patti, G. J. (2017). Systems-Level Annotation of a Metabolomics Data Set Reduces 25 000 Features to Fewer than 1000 Unique Metabolites. Analytical Chemistry, 89(19), 10397–10406. **https://doi.org/10.1021/acs.analchem.7b02380**.

142 Hill, N. S., Buske, P. J., Shi, Y., & Levin, P. A. (2013). A Moonlighting Enzyme Links Escherichia coli Cell Size with Central Metabolism. PLoS Genetics, 9(7). **https://doi.org/10.1371/journal.pgen.1003663**.

143 Willyard, C. (2018). New human gene tally reignites debate. Nature, 558(7710), 354–355. **https://doi.org/10.1038/d41586-018-05462-w**.

144 Gibson, D. G., Glass, J. I., Lartigue, C., Noskov, V. N., Chuang, R. Y., Algire, M. A., ... Venter, J. C. (2010). Creation of a bacterial cell controlled by a chemically synthesized genome. Science, 329(5987), 52–56. **https://doi.org/10.1126/science.1190719**.

145 Westberg, J., Persson, A., Holmberg, A., Goesmann, A., Lundeberg, J., Johansson, K. E., ... Uhlén, M. (2004). The genome sequence of Mycoplasma mycoides subsp. mycoides SC type strain PG1T, the causative agent of contagious bovine pleuropneumonia (CBPP). Genome Research, 14(2), 221–227. **https://doi.org/10.1101/gr.1673304**.

146 Hutchison, C. A., Chuang, R. Y., Noskov, V. N., Assad-Garcia, N., Deerinck, T. J., Ellisman, M. H., ... Venter, J. C. (2016). Design and synthesis of a minimal bacterial genome. Science, 351(6280). **https://doi.org/10.1126/science.aad6253**.

147 Salis, H. M. (2011). The ribosome binding site calculator. In C. Voigt (Ed.), Methods in Enzymology (Vol. 498, pp. 19–42). **https://doi.org/10.1016/B978-0-12-385120-8.00002-4.**

148 Hillson, N. J., Rosengarten, R. D., & Keasling, J. D. (2012). j5 DNA assembly design automation software. ACS Synthetic Biology, 1(1), 14–21. **https://doi.org/10.1021/sb2000116.**

149 Flydal, M. I., Alcorlo-Pagés, M., Johannessen, F. G., Martínez-Caballero, S., Skjærven, L., Fernandez-Leiro, R., ... Hermoso, J. A. (2019). Structure of full-length human phenylalanine hydroxylase in complex with tetrahydrobiopterin. Proceedings of the National Academy of Sciences of the United States of America, 166(23), 11229–11234. **https://doi.org/10.1073/pnas.1902639116.**

Jaffe, E. K. (2017). New protein structures provide an updated understanding of phenylketonuria. Molecular Genetics and Metabolism, 121(4), 289–296. **https://doi.org/10.1016/j.ymgme.2017.06.005.**

150 KYAT1 - Kynurenine--Oxoglutarate Transaminase 1 - Homo Sapiens (Human) - KYAT1 Gene & Protein. UniProt. **https://www.uniprot.org/uniprot/Q16773** (Accessed December 24, 2019).

Han, Q., Robinson, H., Cai, T., Tagle, D. A., & Li, J. (2009). Structural insight into the inhibition of human kynurenine aminotransferase I/Glutamine transaminase K. Journal of Medicinal Chemistry, 52(9), 2786–2793. **https://doi.org/10.1021/jm9000874.**

151 Blau, N., Van Spronsen, F. J., & Levy, H. L. (2010). Phenylketonuria. The Lancet, 376(9750), 1417–1427. **https://doi.org/10.1016/S0140-6736(10)60961-0.**

152 Pline-Srnic, W. (2005). Technical performance of some commercial glyphosate-resistant crops. Pest Management Science, 61(3), 225–234. **https://doi.org/10.1002/ps.1009.**

153 Ye, X., & Beyer, P. (2000). Engineering the provitamin A (β-carotene) biosynthetic pathway into (carotenoid-free) rice endosperm. Science, 287(5451), 303–305. **https://doi.org/10.1126/science.287.5451.303.**

154 Paine, J. A., Shipton, C. A., Chaggar, S., Howells, R. M., Kennedy, M. J., Vernon, G., ... Drake, R. (2005). Improving the nutritional value of Golden Rice through increased pro-vitamin A content. Nature Biotechnology, 23(4), 482–487. **https://doi.org/10.1038/nbt1082.**

155 Enserink, M. (2008). Tough Lessons From Golden Rice. Science, 320(5875), 468–471. **https://doi.org/10.1126/science.320.5875.468.**

156 Genetic Engineering. Greenpeace International. **http://www.greenpeace.org/international/en/campaigns/agriculture/problem/genetic-engineering/** (Accessed June 16, 2019).

157 Powell, W. (2016). New Genetically Engineered American Chestnut Will Help Restore the Decimated, Iconic Tree. The Conversation. **http://theconversation.com/new-genetically-engineered-american-chestnut-will-help-restore-the-decimated-iconic-tree-52191** (Accessed January 19, 2020).

158 NRC: Regulations Title 10, Code of Federal Regulations. United States Nuclear Regulatory Commission. **https://www.nrc.gov/reading-rm/doc-collections/cfr/** (Accessed December 24, 2019).

159 Enserink, M. (2008). Tough Lessons From Golden Rice. Science, 320(5875), 468–471. **https://doi.org/10.1126/science.320.5875.468.**

Mandell, D. J., Lajoie, M. J., Mee, M. T., Takeuchi, R., Kuznetsov, G., Norville, J. E., ... Church, G. M. (2015). Biocontainment of genetically modified organisms by synthetic protein design. Nature, 518(7537), 55–60. **https://doi.org/10.1038/nature14121.**

160 Anderson, L. (2014). Why Does Everyone Hate Monsanto? Modern Farmer. **https://modernfarmer.com/2014/03/monsantos-good-bad-pr-problem/** (Accessed January 14, 2021).

161 Monsanto Legal Cases. Wikipedia. **https://en.wikipedia.org/w/index.php?title=Monsanto_legal_cases&oldid=1012233322.** (Accessed March 15, 2021)

162 Brothers, W. (2020). A Timeline of COVID-19 Vaccine Development. BioSpace. **https://www.biospace.com/article/a-timeline-of-covid-19-vaccine-development/** (Accessed December 18, 2020).

163 Regalado, A. (2020). What Are the Ingredients of Pfizer's Covid-19 Vaccine? MIT Technology Review. **https://www.technologyreview.com/2020/12/09/1013538/what-are-the-ingredients-of-pfizers-covid-19-vaccine/** (Accessed December 18, 2020).

Chow, D. (2020). What is mRNA? How Pfizer and Moderna tapped new tech to make coronavirus vaccines. NBC News. **https://www.nbcnews.com/science/science-news/what-mrna-how-pfizer-moderna-tapped-new-tech-make-coronavirus-n1248054** (Accessed December 18, 2020).

164 Wallace-Wells, D. (2020). We Had the Vaccine the Whole Time. Intelligencer. **https://nymag.com/intelligencer/2020/12/moderna-covid-19-vaccine-design.html** (Accessed December 18, 2020).

165 How Influenza (Flu) Vaccines Are Made. (2020). Center for Disease Control. **https://www.cdc.gov/flu/prevent/how-fluvaccine-made.htm** (Accessed December 18, 2020).

166 Yeung, J. (2020). The US keeps millions of chickens in secret farms to make flu vaccines. But their eggs won't work for coronavirus. CNN. **https://www.cnn.com/2020/03/27/health/chicken-egg-flu-vaccine-intl-hnk-scli/index.html** (Accessed December 18, 2020).

Influenza Vaccine Manufacturing Process. (2015). GlaxoSmithKline.

https://www.gskvaccination.com/content/dam/assets/516007R1_FluVaccineManufacturingProcess.pdf (Accessed December 18, 2020).

167 Branton, H., Liddell, J., Rosamonte, M., Taylor, L., Lucas, C., & Humphreys, J. (2020, April 30). Teaming up for Vaccines. The Chemical Engineer. **https://www.thechemicalengineer.com/features/teaming-up-for-vaccines/** (Accessed December 18, 2020).

168 Nora, L. C., Westmann, C. A., Martins-Santana, L., Alves, L. de F., Monteiro, L. M. O., Guazzaroni, M. E., & Silva-Rocha, R. (2019). The art of vector engineering: towards the construction of next-generation genetic tools. Microbial Biotechnology, 12(1), 125–147. **https://doi.org/10.1111/1751-7915.13318.**

169 Baumgart, A. K., & Beyer, M. (2017). Genetic engineering as a tool for the generation of mouse models to understand disease phenotypes and gene function. Current Opinion in Biotechnology, 48, 228–233. **https://doi.org/10.1016/j.copbio.2017.06.012.**

170 Filipiak, W. E., Hughes, E. D., Gavrilina, G. B., LaForest, A. K., & Saunders, T. L. (2019). Next Generation Transgenic Rat Model Production. Rat Genomics, 97–114. **https://doi.org/10.1007/978-1-4939-9581-3_4**

171 Wang, Y., Huang, J., & Zhao, J. (2017). Gene engineering in swine for agriculture. Journal of Integrative Agriculture, 16(12), 2792–2804. **https://doi.org/10.1016/S2095-3119(17)61766-0.**

172 List of Animals That Have Been Cloned. (2020). Wikipedia. **https://en.wikipedia.org/w/index.php?title=List_of_animals_that_have_been_cloned&oldid=984765213** (Accessed December 18, 2020).

173 Lee, A. (2020). A California Couple Was Heartbroken to Say Goodbye to Their Beloved Dog, Marley. So They Cloned Him. CNN. **https://www.cnn.com/2020/02/27/us/california-family-clones-dog-trnd/index.html** (Accessed December 19, 2020).

174 GMOs and Veganism - What Are GMOs the Pros and Cons. (2018). My Vegan Dreams. **https://myvegandreams.com/gmos-and-veganism/** (January 18, 2020).

175 Anchel, D. (2016). Methods and Compositions for Egg White Protein Production. **https://patents.google.com/patent/WO2016077457A1/en** (Accessed December 19, 2020).

CHAPTER 6

176 Mitchell, N. (2017). The Diet Book Industry Is a Lie – and We've All Been Sucked into It. The Telegraph. **https://www.telegraph.co.uk/health-fitness/body/the-diet-book-industry-is-a-lie--and-weve-all-been-sucked-into-i/** (Accessed April 18, 2020).

177 Pollan, M. (2008). In Defence of Food: The Myth of Nutrition and the Pleasures of Eating. Penguin.

178 Smil, V. (2013). Should We Eat Meat?: Evolution and Consequences of Modern Carnivory (1st edition). Wiley-Blackwell.

179 Masola, B., & Ngubane, N. P. (2010). The activity of phosphate-dependent glutaminase from the rat small intestine is modulated by ADP and is dependent on integrity of mitochondria. Archives of Biochemistry and Biophysics, 504(2), 197–203. **https://doi.org/10.1016/j.abb.2010.09.002.**

180 Suárez, I., Bodega, G., & Fernández, B. (2002). Glutamine synthetase in brain: Effect of ammonia. Neurochemistry International, 41(2–3), 123–142. **https://doi.org/10.1016/S0197-0186(02)00033-5.**

181 Facilitated via the Glutamate dehydrogenase enzyme.

182 I adapted this figure, available under the Creative Commons BY-SA 3.0 license. Original figure: Wikipedia users: **Narayanese, WikiUserPedia, YassineMrabet, TotoBaggins** "File:Citric Acid Cycle Noi.Svg." 2012. Wikipedia. **https://en.wikipedia.org/w/index.php?title=File:Citric_acid_cycle_noi.svg&oldid=475143008** (Accessed January 6, 2020).

183 I adapted this figure, available under the Creative Commons BY-SA 4.0 license. Original figure: Wikipedia users: Shannon1. "Mississippi River." 2019. Wikipedia. **https://en.wikipedia.org/w/index.php?title=Mississippi_River&oldid=933292254** (Accessed January 4, 2020).

184 Sekar, K., Linker, S. M., Nguyen, J., Grünhagen, A., Stocker, R., & Sauer, U. (2020). Bacterial Glycogen Provides Short-Term Benefits in Changing Environments. Applied and Environmental Microbiology, 86(9), 1–12. **https://doi.org/10.1128/AEM.00049-20.**

Weber, C. A., Sekar, K., Tang, J. H., Warmer, P., Sauer, U., & Weis, K. (2020). β-Oxidation and autophagy are critical energy providers during acute glucose depletion in Saccharomyces cerevisiae. Proceedings of the National Academy of Sciences of the United States of America, 117(22). **https://doi.org/10.1073/pnas.1913370117.**

Zampieri, M., Sekar, K., Zamboni, N., & Sauer, U. (2017). Frontiers of high-throughput metabolomics. Current Opinion in Chemical Biology, 36. **https://doi.org/10.1016/j.cbpa.2016.12.006.**

185 Figure adapted from Sekar, K., Rusconi, R., Sauls, J. T., Fuhrer, T., Noor, E., Nguyen, J., . . . Sauer, U. (2018). Synthesis and degradation of FtsZ quantitatively predict the first cell division in starved bacteria. Molecular Systems Biology, 14(11), 8623. **https://doi.org/10.15252/msb.20188623.**

186 Collagen Market Size, Share & Trends Analysis Report By Source, By Product (Gelatin, Hydrolyzed, Native, Synthetic), By Application (Food & Beverages, Healthcare, Cosmetics), By Region, And Segment Forecasts, 2021 - 2028. Grand View Research. **https://www.grandviewresearch.com/industry-analysis/collagen-market** (Accessed December 19, 2020).

187 Schwartz, S. R., & Park, J. (2012). Ingestion of BioCell Collagen®, a novel hydrolyzed chicken sternal cartilage extract; enhanced blood microcirculation and reduced facial aging signs. Clinical Interventions in Aging, 7, 267–273. **https://doi.org/10.2147/CIA.S32836.**

Proksch, E., Segger, D., Degwert, J., Schunck, M., Zague, V., & Oesser, S. (2013). Oral supplementation of specific collagen peptides has beneficial effects on human skin physiology: A double-blind, placebo-controlled study. Skin Pharmacology and Physiology, 27(1), 47–55. **https://doi.org/10.1159/000351376.**

Zdzieblik, D., Oesser, S., Baumstark, M. W., Gollhofer, A., & König, D. (2015). Collagen peptide supplementation in combination with resistance training improves body composition and increases muscle strength in elderly sarcopenic men: A randomised controlled trial. British Journal of Nutrition, 114(8), 1237–1245. **https://doi.org/10.1017/S0007114515002810.**

188 Marshall, L. (2019). Collagen: 'Fountain of Youth' or Edible Hoax? WebMD. **https://www.webmd.com/skin-problems-and-treatments/news/20191212/collagen-supplements-what-the-research-shows** (Accessed December 19, 2020).

189 McRae, M. (2018). Eating Collagen Is Becoming a Health Fad, And We Can Only Sit Back And Sigh. ScienceAlert. **https://www.sciencealert.com/science-behind-collagen-supplement-market-increase** (Accessed December 19, 2020).

190 Drouin, G., Godin, J.-R., & Page, B. (2011). The Genetics of Vitamin C Loss in Vertebrates. Current Genomics, 12(5), 371–378. **https://doi.org/10.2174/138920211796429736.**

191 Essential Amino Acid. (2020). Wikipedia. **https://en.wikipedia.org/w/index.php?title=Essential_amino_acid&oldid=934064805** (Accessed January 4, 2020).

192 David, L. A., Maurice, C. F., Carmody, R. N., Gootenberg, D. B., Button, J. E., Wolfe, B. E., ... Turnbaugh, P. J. (2014). Diet Rapidly Alters the Human Gut Microbiota. Nature, 505(7484), 559–563. **https://doi.org/10.1038/nature12820.**

193 High-Protein, Low-Carb Diets Explained. WebMD. **https://www.webmd.com/diet/guide/high-protein-low-carbohydrate-diets.** (Accessed March 28, 2021).

194 Britten, R. J. (2002). Divergence between samples of chimpanzee and human DNA sequences is 5%, counting indels. Proceedings of the National Academy of Sciences of the United States of America, 99(21), 13633–13635. **https://doi.org/10.1073/pnas.172510699.**

195 Armelagos, G. J. (2014). Brain Evolution, the Determinates of Food Choice, and the Omnivore's Dilemma. Critical Reviews in Food Science and Nutrition, 54(10), 1330–1341. **https://doi.org/10.1080/10408398.2011.635817.**

196 Wanjek, C. (2012). Meat, Cooked Foods Needed for Early Human Brain. Live Science. **https://www.livescience.com/24875-meat-human-brain.html** (Accessed November 15, 2020).

197 Cohen, Baruch C. The Ethics Of Using Medical Data From Nazi Experiments. Jewish Law. **http://www.jlaw.com/Articles/NaziMedEx.html** (Accessed January 6, 2020).

198 Infrastructure. (2018). Toppr-guides. **https://www.toppr.com/guides/economics/indian-economy-on-the-eve-of-independence/infrastructure/** (Accessed January 6, 2020).

199 Mizushima, N., & Klionsky, D. J. (2007). Protein turnover via autophagy: Implications for metabolism. Annual Review of Nutrition, 27, 19–40. **https://doi.org/10.1146/annurev.nutr.27.061406.093749.**

200 Waterlow, J. C. (1975). Protein turnover in the whole body. Nature, 253(5488), 157–157. **https://doi.org/10.1038/253157a0.**

201 To get this number: The empirical formula for proteins is $C_4H_{6.2}N_1O_{1.2}P_{0.01}S_{0.01}$, which translates to 88 grams per mole. 1 mole of nitrogen is 14 grams, so 88/14*4 = 25. Formula from:

Perrett, D. (2007). From "protein" to the beginnings of clinical proteomics. Proteomics. Clinical Applications, 1(8), 720–738. **https://doi.org/10.1002/prca.200700525.**

202 Rice, White, Glutinous, Cooked Nutrition Facts & Calories. NutritionData. **https://nutritiondata.self.com/facts/cereal-grains-and-pasta/5722/2** (Accessed January 19, 2020).

203 Lemon, P. W. R., Tarnopolsky, M. A., MacDougall, J. D., & Atkinson, S. A. (1992). Protein requirements and muscle mass/strength changes during intensive training in novice bodybuilders. Journal of Applied Physiology, 73(2), 767–775. **https://doi.org/10.1152/jappl.1992.73.2.767.**

204 Snyder W.S., Cook M.J., Nasset E.S., Karhausen L.R., Howells G.P., Tipton I.H. (1975). Report of the Task Group on Reference Man. Annals of the ICRP/ICRP Publication. 23(1). **https://doi.org/10.1016/S0074-2740(75)80015-8.**

205 Short, K. R., Vittone, J. L., Bigelow, M. L., Proctor, D. N., & Nair, K. S. (2004). Age and aerobic exercise training effects on whole body and muscle protein metabolism. American Journal of Physiology - Endocrinology and Metabolism, 286(1 49-1), 92–101. **https://doi.org/10.1152/ajpendo.00366.2003.**

206 Mitchell, H., Hamilton, T., Steggerda, F., & Bean, H. (1945). THE CHEMICAL COMPOSITION OF THE ADULT HUMAN BODY AND ITS BEARING ON THE BIOCHEMISTRY OF GROWTH. Journal of Biological Chemistry, 158(3), 625–637. **https://doi.org/10.1016/S0021-9258(19)51339-4.**

207 Nilsson, A., Mardinoglu, A., & Nielsen, J. (2017). Predicting growth of the healthy infant using a genome scale metabolic model. Npj Systems Biology and Applications, 3(1), 1–8. **https://doi.org/10.1038/s41540-017-0004-5.**

208 Protein: Are You Getting Enough? WebMD. **https://www.webmd.com/food-recipes/protein** (Accessed January 20, 2020).

Nutrient Recommendations : Dietary Reference Intakes (DRI). National Institute of Health. **https://ods.od.nih.gov/Health_Information/Dietary_Reference_Intakes.aspx** (Accessed June 6, 2020).

209 Nutrient Recommendations: Dietary Reference Intakes (DRI). **https://ods.od.nih.gov/Health_Information/Dietary_Reference_Intakes.aspx** (June 6, 2020).

210 Trumbo, P., Schlicker, S., Yates, A. A., Poos, M., & Food and Nutrition Board of the Institute of Medicine, The National Academies. (2002). Dietary reference intakes for energy, carbohydrate, fiber, fat, fatty acids, cholesterol, protein and amino acids. Journal of the American Dietetic Association, 102(11), 1621–1630. **https://doi.org/10.1016/s0002-8223(02)90346-9.**

211 Fulgoni, V. L. (2008). Current protein intake in America: Analysis of the National Health and Nutrition Examination Survey, 2003-2004. American Journal of Clinical Nutrition, 87(5), 2003–2004. **https://doi.org/10.1093/ajcn/87.5.1554s.**

212 Jansen, G. R., & Howe, E. E. (1964). WORLD PROBLEMS IN PROTEIN NUTRITION. The American Journal of Clinical Nutrition, 15(1), 262–274. **https://doi.org/10.1093/ajcn/15.5.262.**

213 The Personalized Nutrition Project - About Us. The Personalized Nutrition Project. **http://personalnutrition.org/AboutGuests.aspx** (Accessed January 5, 2020).

214 Zeevi, D., Korem, T., Zmora, N., Israeli, D., Rothschild, D., Weinberger, A., ... Segal, E. (2015). Personalized Nutrition by Prediction of Glycemic Responses. Cell, 163(5), 1079–1094. **https://doi.org/10.1016/j.cell.2015.11.001.**

215 Segal, E. "Personalized medicine approaches based on gut microbiota" Lecture. SystemsX.ch 2017 Conference. September 6, 2017. Zurich, Switzerland.

216 Kalsbeek, A., La Fleur, S., & Fliers, E. (2014). Circadian control of glucose metabolism. Molecular Metabolism, 3(4), 372–383. **https://doi.org/10.1016/j.molmet.2014.03.002.**

217 Adibi, S. A., Gray, S. J., & Menden, E. (1967). The kinetics of amino acid absorption and alteration of plasma composition of free amino acids after intestinal perfusion of amino acid mixtures. The American Journal of Clinical Nutrition, 20(1), 24–33. **https://doi.org/10.1093/ajcn/20.1.24.**

218 Adibi, S. A., & Mercer, D. W. (1973). Protein digestion in human intestine as reflected in luminal, mucosal, and plasma amino acid concentrations after meals. Journal of Clinical Investigation, 52(7), 1586–1594. **https://doi.org/10.1172/JCI107335.**

219 Luo, Q., Chen, D., Boom, R. M., & Janssen, A. E. M. (2018). Revisiting the enzymatic kinetics of pepsin using isothermal titration calorimetry. Food Chemistry, 268(February), 94–100. **https://doi.org/10.1016/j.foodchem.2018.06.042.**

220 Verkempinck, S. H. E., Salvia-Trujillo, L., Infantes Garcia, M. R., Hendrickx, M. E., & Grauwet, T. (2019). From single to multiresponse modelling of food digestion kinetics: The case of lipid digestion. Journal of Food Engineering, 260(May), 40–49. **https://doi.org/10.1016/j.jfoodeng.2019.04.018.**

221 Hanigan, M. D., & Daley, V. L. (2020). Use of Mechanistic Nutrition Models to Identify Sustainable Food Animal Production. Annual Review of Animal Biosciences, 8(1). **https://doi.org/10.1146/annurev-animal-021419-083913.**

222 Luo, Q., Chen, D., Boom, R. M., & Janssen, A. E. M. (2018). Revisiting the enzymatic kinetics of pepsin using isothermal titration calorimetry. Food Chemistry, 268(February), 94–100. **https://doi.org/10.1016/j.foodchem.2018.06.042.**

223 Baum, J., & Wolfe, R. (2015). The Link between Dietary Protein Intake, Skeletal Muscle Function and Health in Older Adults. Healthcare, 3(3), 529–543. **https://doi.org/10.3390/healthcare3030529.**

224 Curley, K. (2018). What Is Enriched Bread? SF Gate. **https://healthyeating.sfgate.com/enriched-bread-2900.html** (Accessed December 20, 2020).

225 Leung, A. M., Braverman, L. E., & Pearce, E. N. (2012). History of U.S. iodine fortification and supplementation. Nutrients, 4(11), 1740–1746. **https://doi.org/10.3390/nu4111740.**

226 Hollowell, J. G., Staehling, N. W., Hannon, W. H., Flanders, D. W., Gunter, E. W., Maberly, G. F., ... Jackson, R. J. (1998). Iodine nutrition in the United States. Trends and public health implications: Iodine excretion data from national Health and Nutrition Examination Surveys I and III (1971-1974 and 1988-1994). Journal of Clinical Endocrinology and Metabolism, 83(10), 3401–3408. **https://doi.org/10.1210/jcem.83.10.5168.**

227 Rowland, E., Dugbaza, J., & House, B. (2010). Consumer Awareness, Attitudes and Behaviours to Fortified Foods. Food Standards Australia New Zealand. **https://www.foodstandards.gov.au/publications/Pages/Consumer-awareness-attitudes-behaviours-to-fortified-foods---qualitative-.aspx**

228 Allen, R. H., Seetharam, B., Podell, E., & Alpers, D. H. (1978). Effect of Proteolytic Enzymes on the Binding of Cobalamin to R Protein and Intrinsic Factor. Journal of Clinical Investigation, 61(1), 47–54. **https://doi.org/10.1172/jci108924.**

229 Martens, J. H., Barg, H., Warren, M., & Jahn, D. (2002). Microbial production of vitamin B12. Applied Microbiology and Biotechnology, 58(3), 275–285. **https://doi.org/10.1007/s00253-001-0902-7.**

230 Wang, H., Li, L., Qin, L. L., Song, Y., Vidal-Alaball, J., & Liu, T. H. (2018). Oral vitamin B12 versus intramuscular vitamin B12 for vitamin B12 deficiency. Cochrane Database of Systematic Reviews, 2018(3). **https://doi.org/10.1002/14651858.CD004655.pub3.**

231 Butler, C. C., Vidal-Alaball, J., Cannings-John, R., McCaddon, A., Hood, K., Papaioannou, A., ... Goringe, A. (2006). Oral vitamin B12 versus intramuscular vitamin B12 for vitamin B12 deficiency: a systematic review of randomized controlled trials. Family Practice, 23(3), 279–285. **https://doi.org/10.1093/fampra/cml008.**

232 Bernhoft, R. A. (2012). Mercury toxicity and treatment: A review of the literature. Journal of Environmental and Public Health, 2012. **https://doi.org/10.1155/2012/460508.**

233 Alexander, D. D., Weed, D. L., Miller, P. E., & Mohamed, M. A. (2015). Red Meat and Colorectal Cancer: A Quantitative Update on the State of the Epidemiologic Science. Journal of the American College of Nutrition, 34(6), 521–543. **https://doi.org/10.1080/07315724.2014.992553.**

234 Richi, E. B., Baumer, B., Conrad, B., Darioli, R., Schmid, A., & Keller, U. (2015). Health risks associated with meat consumption: A review of epidemiological studies. International Journal for Vitamin and Nutrition Research, 85(1–2), 70–78. **https://doi.org/10.1024/0300-9831/a000224.**

235 Zywica, M. (2004). Olestra. OU Kosher. **https://oukosher.org/blog/industrial-kosher/olestra/** (Accessed January 28, 2020).

236 Victory, J. (2007). Intestinal Cramp Complaints Crimp 'Light' Chip Labels. ABC News. **https://abcnews.go.com/Health/story?id=2028655&page=1** (Accessed March 22, 2020).

237 Sandler, R. S., Zorich, N. L., Filloon, T. G., Wiseman, H. B., Lietz, D. J., Brock, M. H., Royer, M. G., & Miday, R. K. (1999). Gastrointestinal Symptoms in 3181 Volunteers Ingesting Snack Foods Containing Olestra or Triglycerides for 6 Weeks. A Randomized, Placebo-Controlled Trial. Annals of Internal Medicine, 130(4 Pt 1), 253–261. **https://doi.org/10.7326/0003-4819-130-4_part_1-199902160-00002.**

238 Swithers, S. E., Ogden, S. B., & Davidson, T. L. (2011). Fat substitutes promote weight gain in rats consuming high-fat diets. Behavioral Neuroscience, 125(4), 512–518. **https://doi.org/10.1037/a0024404.**

239 Swithers, S. E. (2013). Artificial sweeteners produce the counterintuitive effect of inducing metabolic derangements. Trends in Endocrinology and Metabolism, 24(9), 431–441. **https://doi.org/10.1016/j.tem.2013.05.005.**

240 Diether, M., & Sauer, U. (2017). Towards detecting regulatory protein–metabolite interactions. Current Opinion in Microbiology, 39, 16–23. **https://doi.org/10.1016/j.mib.2017.07.006.**

CHAPTER 7

241 Helsinki, M. (2013). Fazer Salmiakki Ice Cream. **https://vimeo.com/62767130** (Accessed January 29, 2020).

242 Blum, K., Cshen, A. L. C., Giordano, J., Borsten, J., Chen, T. J. H., Hauser, M., Simpatico, T., Femino, J., Braverman, E. R., & Barh, D. (2012). The addictive brain: All roads lead to dopamine. Journal of Psychoactive Drugs, 44(2), 134–143. **https://doi.org/10.1080/02791072.2012.685407.**

243 Burger, K. S., & Stice, E. (2012). Frequent ice cream consumption is associated with reduced striatal response to receipt of an ice cream-based milkshake. American Journal of Clinical Nutrition, 95(4), 810–817. **https://doi.org/10.3945/ajcn.111.027003.**

244 Lucas, A. (2019). 5 Charts That Show How Milk Sales Changed and Made It Tough for Dean Foods to Avert Bankruptcy. CNBC. **https://www.cnbc.com/2019/11/13/5-charts-that-show-how-milk-sales-have-changed.html** (Accessed December 24, 2020).

245 U.S. retail market data for the plant-based industry - New SPINS Retail Sales Data. (2018). The Good Food Institute. **https://www.gfi.org/marketresearch** (Accessed December 25, 2020).

246 Bartolotto, C. (2015). Does Consuming Sugar and Artificial Sweeteners Change Taste Preferences? The Permanente Journal, 19(3), 81–84. **https://doi.org/10.7812/TPP/14-229.**

Palacios, O. M., Badran, J., Spence, L., Drake, M. A., Reisner, M., & Moskowitz, H. R. (2010). Measuring Acceptance of Milk and Milk Substitutes Among Younger and Older Children. Journal of Food Science, 75(9), 522–526. **https://doi.org/10.1111/j.1750-3841.2010.01839.x.**

Lawrence, S. E., Lopetcharat, K., & Drake, M. A. (2016). Preference Mapping of Soymilk with Different U.S. Consumers. Journal of Food Science, 81(2), S463–S476. **https://doi.org/10.1111/1750-3841.13182.**

247 Diamond, J. M. (1999). Guns, Germs, and Steel: The Fates of Human Societies (1st edition). W. W. Norton & Company.

248 Sánchez-Pérez, R., Pavan, S., Mazzeo, R., Moldovan, C., Aiese Cigliano, R., Del Cueto, J., ... Lindberg Møller, B. (2019). Mutation of a bHLH transcription factor allowed almond domestication. Science, 364(6445), 1095–1098. **https://doi.org/10.1126/science.aav8197.**

249 Sapolsky, R. M. (2018). Behave: The Biology of Humans at Our Best and Worst (Illustrated edition). Penguin Books.

250 Bromberg-Martin, E. S., Matsumoto, M., & Hikosaka, O. (2010). Dopamine in Motivational Control: Rewarding, Aversive, and Alerting. Neuron, 68(5), 815–834. **https://doi.org/10.1016/j.neuron.2010.11.022.**

251 Robinson, D. L., Phillips, P. E. M., Budygin, E. A., Trafton, B. J., Garris, P. A., & Wightman, R. M. (2001). Sub-second changes in accumbal dopamine during sexual behavior in male rats. NeuroReport, 12(11), 2549–2552. **https://doi.org/10.1097/00001756-200108080-00051.**

252 Bromberg-Martin, E. S., Matsumoto, M., & Hikosaka, O. (2010). Dopamine in Motivational Control: Rewarding, Aversive, and Alerting. Neuron, 68(5), 815–834. **https://doi.org/10.1016/j.neuron.2010.11.022.**

253 Covington, P., Adams, J., & Sargin, E. (2016). Deep Neural Networks for YouTube Recommendations. RecSys 2016 - Proceedings of the 10th ACM Conference on Recommender Systems, 191–198. **https://doi.org/10.1145/2959100.2959190.**

254 Solsman, J. E. (2018). YouTube's AI is the puppet master over most of what you watch. CNET. **https://www.cnet.com/news/youtube-ces-2018-neal-mohan/** (Accessed December 25, 2020).

255 Wise, R. A. (2008). Dopamine and reward: The anhedonia hypothesis 30 years on. Neurotoxicity Research, 14(2–3), 169–183. **https://doi.org/10.1007/BF03033808.**

256 Sapolsky, R. M. (2018). Behave: The Biology of Humans at Our Best and Worst (Illustrated edition). Penguin Books.

257 Schultz, W. (2010). Dopamine signals for reward value and risk: basic and recent data. Behavioral and Brain Functions: BBF, 6, 24. **https://doi.org/10.1186/1744-9081-6-24**.

258 Gilbert, D. (2007). Stumbling on Happiness. Vintage.

259 Kahneman, D., & Deaton, A. (2010). High income improves evaluation of life but not emotional well-being. Proceedings of the National Academy of Sciences of the United States of America, 107(38), 16489–16493. **https://doi.org/10.1073/pnas.1011492107**.

Stevenson, B. & Wolfers, J. (2015). Subjective well-being and income. National Bureau of Economic Research, 1. **https://doi.org//10.3386/w18992**.

260 Buettner, A. (2017). Springer Handbook of Odor. In A. Buettner (Ed.), Springer Handbooks. **https://doi.org/10.1007/978-3-319-26932-0**.

261 Inside the Strange Science of the Fake Meat That 'Bleeds.' Wired. **https://www.wired.com/story/the-impossible-burger/** (March 29, 2020).

262 Liz Specht on The Future of Food, Alternative Meats, and Many Materials. What's Now: San Francisco. **https://www.youtube.com/watch?v=poBiQnY3y9k&feature=emb_logo** (March 29, 2020).

263 Metzinger, T. (2009). The Ego Tunnel: The Science of the Mind and the Myth of the Self (1st edition). Basic Books.

264 Pinker, S. (2009). How the Mind Works (Illustrated edition). W. W. Norton & Company.

265 Penfield, W., & Boldrey, E. (1937). Somatic Motor and Sensory Representation in the Cerebral Cortex of Man as Studied by Electrical Stimulation. Brain, 389–443. **https://doi.org/10.1093/brain/60.4.389**.

266 Sanes, J., Donoghue, J., Thangaraj, V., Edelman, R., & Warach, S. (1995). Shared neural substrates controlling hand movements in human motor cortex. Science, 268(5218), 1775–1777. **https://doi.org/10.1126/science.7792606.**

267 Lienhard, D. A. (2017). Roger Sperry's Split Brain Experiments (1959–1968). Embryo Project Encyclopedia. **http://embryo.asu.edu/handle/10776/13035.**

268 Harris, S. (2014). Waking Up: A Guide to Spirituality Without Religion (Reprint edition). Simon & Schuster.

269 Norretranders, T. (1999). The User Illusion: Cutting Consciousness Down to Size (1st edition). Penguin Books.

270 Harris, S. (2014). Waking Up: A Guide to Spirituality Without Religion (Reprint edition). Simon & Schuster.

Hameroff, S. R., Kaszniak, A. W., & Scott, A. C. (Eds.). (1996). Toward a Science of Consciousness: The First Tucson Discussions and Debates. A Bradford Book.

271 Is Buddhism True? Making Sense Podcast. **https://samharris.org/podcasts/is-buddhism-true/.**

272 Cepelewicz, J. (2017). Is Consciousness Fractal? Nautilus. **http://nautil.us/issue/47/consciousness/is-consciousness-fractal** (Accessed May 31, 2020).

273 "Blind Spot (Vision)." (2020). Wikipedia. **https://en.wikipedia.org/w/index.php?title=Blind_spot_(vision)&oldid=939210050** (Accessed March 30, 2020).

274 Gowin, J. 7 Reasons We Can't Turn Down Fast Food. Psychology Today. **http://www.psychologytoday.com/blog/you-illuminated/201108/7-reasons-we-cant-turn-down-fast-food** (Accessed December 27, 2020).

275 Managing Pain after Burn Injury. Model Systems Knowledge Translation Center (MSKTC). **https://msktc.org/burn/factsheets/Managing-Pain-After-Burn-Injury** (Accessed February 15, 2020).

276 Thích Quảng Đức. (2020). Wikipedia. **https://en.wikipedia.org/w/index.php?title=Th%C3%ADch_Qu%E1%BA%A3ng_%C4%90%E1%BB%A9c&oldid=950002550** (Accessed April 25, 2020).

277 Yates, J., Immergut, M., & Graves, J. (2017). The Mind Illuminated: A Complete Meditation Guide Integrating Buddhist Wisdom and Brain Science for Greater Mindfulness. Atria Books.

278 Harari, Y. N. (2017). Homo Deus: A Brief History of Tomorrow (Illustrated edition). Harper.

279 Zabelina, D. L., White, R. A., Tobin, A., & Thompson, L. (2020). The Role of Mindfulness in Viewing and Making Art in Children and Adults. Mindfulness, 11(11), 2604–2612. **https://doi.org/10.1007/s12671-020-01474-8.**

280 Sedlmeier, P., Eberth, J., Schwarz, M., Zimmermann, D., Haarig, F., Jaeger, S., & Kunze, S. (2012). The psychological effects of meditation: A meta-analysis. Psychological Bulletin, 138(6), 1139–1171. **https://doi.org/10.1037/a0028168.**

281 Henriksen, D., Richardson, C., & Shack, K. (2020). Mindfulness and creativity: Implications for thinking and learning. Thinking Skills and Creativity, 37(January), 100689. **https://doi.org/10.1016/j.tsc.2020.100689.**

282 Brown, K. W., Ryan, R. M., & Creswell, J. D. (2007). Mindfulness: Theoretical foundations and evidence for its salutary effects. Psychological Inquiry, 18(4), 211–237. **https://doi.org/10.1080/10478400701598298.**

CHAPTER 8

283 Tomato | Description, Cultivation, & History. Encyclopedia Britannica. **https://www.britannica.com/plant/tomato** (Accessed March 2, 2020).

284 Sen, C. T. (2004). Food Culture in India (Illustrated edition). Greenwood.

285 Mariani, J. F., & Bastianich, L. M. (2011). How Italian Food Conquered the World (1st edition). St. Martin's Press.

286 Hatch, P. J., & Waters, A. (2014). "A Rich Spot of Earth": Thomas Jefferson's Revolutionary Garden at Monticello (Illustrated edition). Yale University Press.

287 The Nestlé Company History. Nestlé Global. **https://www.nestle.com/aboutus/history/nestle-company-history** (Accessed March 2, 2020).

288 Rose, S. (2010). The Great British Tea Heist. Smithsonian Magazine. **https://www.smithsonianmag.com/history/the-great-british-tea-heist-9866709/** (Accessed March 2, 2020).

289 Black, J. (2007). The Trail of Tiramisu. Washington Post. **http://www.washingtonpost.com/wp-dyn/content/article/2007/07/10/AR2007071000327.html** (Accessed March 2, 2020).

290 Yglesias, M. (2012). The Avocado Boom: Brought to You by NAFTA. Slate Magazine. **https://slate.com/business/2012/09/avocado-consumption-is-booming-thanks-to-nafta.html** (Accessed March 2, 2020).

291 Brassica Oleracea. (2020). Wikipedia. **https://en.wikipedia.org/w/index.php?title=Brassica_oleracea&oldid=942209146** (Accessed March 2, 2020).

292 Charles, Dan. (2019). From Culinary Dud To Stud: How Dutch Plant Breeders Built Our Brussels Sprouts Boom. NPR.org. **https://www.npr.org/sections/thesalt/2019/10/30/773457637/from-culinary-dud-to-stud-how-dutch-plant-breeders-built-our-brussels-sprouts-bo** (Accessed March 2, 2020).

293 Rowland, M. P. (2020). Memphis Meats Raises $161 Million In Funding, Aims To Bring Cell-Based Products To Consumers. Forbes. **https://www.forbes.com/sites/michaelpellmanrowland/2020/01/22/memphis-meats-raises-161-million-series-b-funding-round-aims-to-bring-cell-based-products-to-consumers-for-the-first-time/** (Accessed April 25, 2020).

294 O'Brian, M. (2019). How Does the Impossible Burger Look and Taste Like Real Beef? Discover Magazine. **https://www.discovermagazine.com/planet-earth/how-does-the-impossible-burger-look-and-taste-like-real-beef** (Accessed January 10, 2021).

295 Hassan, F. A. M., Abd El-Gawad, M. A. M., & Enab, A. K. (2012). Flavour compounds in cheese (review). International Journal of Academic Research, 4(5), 169–181. **https://doi.org/10.7813/2075-4124.2012/4-5/a.20.**

296 Collins, C. (2016). Kitchen Science: The Chemistry behind Amazing Meringue and Perfect Cappuccino. The Conversation. **http://theconversation.com/kitchen-science-the-chemistry-behind-amazing-meringue-and-perfect-cappuccino-64670** (Accessed January 10, 2021).

297 Rossen, J. (2017). Why Is Trader Joe's Wine Cheaper Than Bottled Water? Mental Floss. **https://www.mentalfloss.com/article/94047/why-trader-joes-wine-cheaper-bottled-water** (Accessed March 7, 2020).

298 Morris, J. R. (2007). Development and commercialization of a complete vineyard mechanization system. HortTechnology, 17(4), 411–420. **https://doi.org/10.21273/horttech.17.4.411.**

299 Titration in Wine Analysis. Lab Manager. **https://www.labmanager.com/product-focus/titration-in-wine-analysis-1373** (Accessed March 7, 2020).

300 Glatter, R. (2012). The Truth Behind The Coconut Water Craze. Forbes. **https://www.forbes.com/sites/robertglatter/2012/08/31/the-truth-behind-the-coconut-water-craze/** (Accessed March 7, 2020).

301 Gibbs, P., & Komitopoulou, E. (2011, May 13). Overview of food preservation technologies. New Food Magazine. **https://www.newfoodmagazine.com/article/4414/overview-of-food-preservation-technologies/** (Accessed March 7, 2020).

302 The Chemistry of Coconut Water. (2016). Coconut Handbook. **https://coconuthandbook.tetrapak.com/chapter/chemistry-coconut-water** (Accessed March 7, 2020).

303 Gordon, A., & Jackson, J. (2017). Case study: Application of appropriate technologies to improve the quality and safety of coconut water. In Food Safety and Quality Systems in Developing Countries (Vol. 2). **https://doi.org/10.1016/B978-0-12-801226-0.00007-4.**

304 Hazel Technologies, Inc. **https://www.hazeltechnologies.com/** (Accessed April 25, 2020).

305 Blankenship, S. M., & Dole, J. M. (2003). 1-Methylcyclopropene: A review. Postharvest Biology and Technology, 28(1), 1–25. **https://doi.org/10.1016/S0925-5214(02)00246-6.**

306 Bellis, M. (2019). The History of Toasters, From Roman Times to Today. ThoughtCo. **https://www.thoughtco.com/history-of-your-toaster-4076981** (Accessed March 8, 2020).

Mitchell, N. (2017, August 25). How A Decade of Domesticity Changed Our Nation's Kitchens. Apartment Therapy. **https://www.apartmenttherapy.com/brief-history-of-1950s-1960s-kitchens-247463** (Accessed March 8, 2020).

307 Banana Ice Cream. (2015). Simple Vegan Blog. **https://simpleveganblog.com/one-ingredient-banana-ice-cream/** (Accessed March 8, 2020).

308 Weetman, R. J., & Gigas, B. (2002). MIXER MECHANICAL DESIGN — FLUID FORCES by Torque Bending Thrust. International Plump Users Symposium, 203–214.

309 Epidemiology Working Group for NCIP Epidemic Response, Chinese Center for Disease Control and Prevention. (2020). [The epidemiological characteristics of an outbreak of 2019 novel coronavirus diseases (COVID-19) in China]. Chinese Journal of Epidemiology, 41(2), 145–151. **https://doi.org/10.3760/cma.j.issn.0254-6450.2020.02.003.**

310 Zhou, P., Yang, X. Lou, Wang, X. G., Hu, B., Zhang, L., Zhang, W., ... Shi, Z. L. (2020). A pneumonia outbreak associated with a new coronavirus of probable bat origin. Nature, 579(7798), 270–273. **https://doi.org/10.1038/s41586-020-2012-7.**

Doucleff, M., & Lohmeyer, S. (2021). WHO Report: Wildlife Farms, Not Market, Likely Source Of Coronavirus Pandemic. NPR.Org. **https://www.npr.org/sections/goatsandsoda/2021/03/29/982272319/who-report-wildlife-farms-not-market-likely-source-of-coronavirus-pandemic** (Accessed April 4, 2021).

311 Menachery, V. D., Yount, B. L., Debbink, K., Agnihothram, S., Gralinski, L. E., Plante, J. A., ... Baric, R. S. (2015). A SARS-like cluster of circulating bat coronaviruses shows potential for human emergence. Nature Medicine, 21(12), 1508–1513. **https://doi.org/10.1038/nm.3985**

312 WHO | FAQs: H5N1 Influenza. World Health Organization. **https://www.who.int/influenza/human_animal_interface/avian_influenza/h5n1_research/faqs/en/** (Accessed April 12, 2020).

313 Thacker, E., & Janke, B. (2008). Swine Influenza Virus: Zoonotic Potential and Vaccination Strategies for the Control of Avian and Swine Influenzas. The Journal of Infectious Diseases, 197(s1), S19–S24. **https://doi.org/10.1086/524988.**

314 Van Boeckel, T. P., Pires, J., Silvester, R., Zhao, C., Song, J., Criscuolo, N. G., . . . Laxminarayan, R. (2019). Global trends in antimicrobial resistance in animals in low- And middle-income countries. Science, 365(6459), 1–55. **https://doi.org/10.1126/science.aaw1944.**

315 Smil, V. (2005). Creating the Twentieth Century: Technical Innovations of 1867-1914 and Their Lasting Impact (Illustrated edition). Oxford University Press.

316 Child, J., Bertholle, L., & Beck, S. (2001). Mastering the Art of French Cooking, Volume I: 50th Anniversary Edition: A Cookbook (40th Anniversary edition). Knopf.

317 Temple, J. (2014). 8 Facts about Julia Child and The French Chef That May Surprise You. IWFS Blog. **https://blog.iwfs.org/2014/09/8-facts-about-julia-child-and-the-french-chef-that-may-surprise-you/** (Accessed January 9, 2021).

318 Cahn, L. (2019). The Most Popular Cooking Show the Year You Were Born. Taste of Home. **https://www.tasteofhome.com/collection/most-popular-cooking-shows/** (Accessed January 9, 2021).

319 DISCOVERY INC.'s HGTV, FOOD NETWORK, TLC AND ID ARE THE TOP NON-NEWS CABLE NETWORKS IN Q2 AMONG W25-54 IN TOTAL DAY. Discovery, Inc. **https://corporate.discovery.com/discovery-newsroom/discovery-inc-s-hgtv-food-network-tlc-and-id-are-the-top-non-news-cable-networks-in-q2-among-w25-54-in-total-day/** (Accessed January 9, 2021).

320 Vittek, S. (2018). The 5 Most Popular Cooking YouTube Channels. Kitchn. **https://www.thekitchn.com/youtube-most-popular-cooking-channels-258119** (Accessed January 9, 2021).

321 The 50 Best Food & Cooking Blogs (Ranked Algorithmically). Detailed. **https://detailed.com/food-blogs/** (Accessed January 9, 2021).

322 About Minimalist Baker. Minimalist Baker. **https://minimalistbaker.com/about/** (Accessed January 9, 2021).

323 Rabb, M. (2021). Gordon Ramsay Says He's 'Turning Vegan' & Shares Eggplant Steak Recipe. The Beet. **https://thebeet.com/gordon-ramsay-says-hes-turning-vegan-and-shares-an-eggplant-steak-recipe/** (Accessed April 4, 2021).

324 CHLOE COSCARELLI. **https://www.chefchloe.com** (Accessed January 9, 2021).

325 Higgins, K. (2018). Makeout Chocolate Chip Cookie Pie. Chocolate Covered Katie. **https://chocolatecoveredkatie.com/makeout-chocolate-chip-cookie-pie/** (Accessed January 9, 2021).

326 The Center For Consumer Freedom Team. (2019). 5 Chemicals Lurking in Plant-Based Meats. Center for Consumer Freedom. **https://www.consumerfreedom.com/2019/05/5-chemicals-lurking-in-plant-based-meats/** (Accessed January 10, 2021).

327 Azeredo, H. M. C., Barud, H., Farinas, C. S., Vasconcellos, V. M., & Claro, A. M. (2019). Bacterial Cellulose as a Raw Material for Food and Food Packaging Applications. Frontiers in Sustainable Food Systems, 3. **https://doi.org/10.3389/fsufs.2019.00007**.

CHAPTER 9

328 Morton, A. (2019). Australia Is Third Largest Exporter of Fossil Fuels behind Russia and Saudi Arabia. The Guardian. **https://www.theguardian.com/environment/2019/aug/19/australia-is-third-largest-exporter-of-fossil-fuels-behind-russia-and-saudi-arabia** (Accessed May 9, 2020).

329 2019 Country Reports on Human Rights Practices: Russia. United States Department of State. **https://www.state.gov/reports/2019-country-reports-on-human-rights-practices/russia/** (Accessed November 27, 2020).

Saudi Arabia | Events of 2019. (2019). Human Rights Watch. **https://www.hrw.org/world-report/2020/country-chapters/saudi-arabia** (Accessed November 27, 2020).

330 Polls reveal citizens' support for climate action and energy transition. (2015). Clean Energy Wire. **https://www.cleanenergywire.org/factsheets/polls-reveal-citizens-support-energiewende** (Accessed May 9, 2020).

331 McCrone, A. (2015). Global Trends in Renewable Energy Investment 2015. The Frankfurt School – UNEP Collaborating Centre for Climate & Sustainable Energy Finance. **https://www.fs-unep-centre.org/research/report.**

332 Teh, C. (2019). $68m Fund to Turn Labs into Food Factories of the Future. The Straits Times. **https://www.straitstimes.com/singapore/68m-fund-to-turn-labs-into-food-factories-of-the-future** (Accessed May 10, 2020).

333 Emissions of Greenhouse Gases in the U.S. (2011). U.S. Energy Information Administration. **https://www.eia.gov/environment/emissions/ghg_report/ghg_methane.php** (Accessed May 10, 2020).

334 Global Warming Potential. (2020). Wikipedia. **https://en.wikipedia.org/w/index.php?title=Global_warming_potential&oldid=951664951** (Accessed May 10, 2020).

335 Brown, T. (2019). China's Pork Crisis Is Bigger than You Think. MarketWatch. **https://www.marketwatch.com/story/chinas-pork-crisis-is-bigger-than-you-think-2019-11-11** (Accessed June 7, 2020).

336 Mackinnon, Jim. (2020). Local Butchers: COVID-19 Impact on Meat Prices May Ease Soon. The Review. **https://www.the-review.com/news/20200602/local-butchers-covid-19-impact-on-meat-prices-may-ease-soon** (Accessed June 7, 2020).

337 Automobile Production, Selected Countries, 1950-2019. (2017). The Geography of Transport Systems. **https://transportgeography.org/?page_id=1343** (Accessed August 3, 2020).

338 León, R. (2019). How SoftBank and Its $100 Billion Vision Fund Has Become a Global Start-up Machine. CNBC. **https://www.cnbc.com/2019/05/17/softbanks-100-billion-vision-fund-reshapes-world-of-venture-capital.html** (Accessed May 10, 2020).

339 Protein Replacement Startups Are Coming for Food Additives as Shiru Launches from Y Combinator. (2019). TechCrunch. **https://social.techcrunch.com/2019/08/16/protein-replacement-startups-are-coming-for-food-additives-as-shiru-launches-from-y-combinator/** (Accessed May 10, 2020).

340 New Protein Fund. Big Idea Ventures. **https://bigideaventures.com/new-protein/** (Accessed May 10, 2020).

341 Founders Fund Backs Its First Food Tech Startup, Hampton Creek Foods, With A $1M Investment. (2013). TechCrunch. **https://social.techcrunch.com/2013/05/20/founders-fund-backs-hampton-creek-foods/** (Accessed May 10, 2020).

342 IndieBio: Creating the Future of Food. CellAgri. **https://www.cell.ag/indiebio-creating-future-of-food/** (Accessed May 10, 2020).

343 Rowland, M. P. (2020, January 22). Memphis Meats Raises $161 Million In Funding, Aims To Bring Cell-Based Products To Consumers. Forbes. **https://www.forbes.com/sites/michaelpellmanrowland/2020/01/22/memphis-meats-raises-161-million-series-b-funding-round-aims-to-bring-cell-based-products-to-consumers-for-the-first-time/** (Accessed May 10, 2020).

344 Shanker, D., Mulvany, L., Hytha, M., & Bloomberg. (2019, May 2). Beyond Meat Just Had the Best IPO of 2019 as Value Soars to $3.8 Billion. Fortune. **https://fortune.com/2019/05/02/beyond-meat-ipo-stock-price/** (Accessed May 10, 2020).

345 Alternative protein research grants. The Good Food Institute. **https://www.gfi.org/researchgrants** (Accessed June 12, 2020).

346 Opportunities. New Harvest. **https://www.new-harvest.org/opportunities** (Accessed June 12, 2020).

347 NSF Award Search: Award#2021132 - GCR: Laying the Scientific and Engineering Foundation for Sustainable Cultivated Meat Production. National Science Foundation. **https://www.nsf.gov/awardsearch/showAward?AWD_ID=2021132&HistoricalAwards=false** (Accessed September 6, 2020).

348 This is based on my personal communication with Good Food Institute organization members during Summer 2020.

349 Gillespie, A. (2019). Materials by Design. NIST. **https://www.nist.gov/featured-stories/materials-design** (Accessed January 11, 2021).

350 Hayden, E. C. (2014). The automated lab. Nature News, 516(7529), 131. **https://doi.org/10.1038/516131a**.

351 Full disclosure: I worked at Emerald Cloud Lab for two years, and I own stock in the company. I went to high school with one of the founders of Culture.

352 Garcia, J. (2020). Tech, Defense Giants Lobbying for Tax Break That Would Save Them Billions. orlandosentinel.com. **https://www.orlandosentinel.com/coronavirus/os-ne-coronavirus-research-tax-break-20200624-smt7czu2vfc6rivz4ypybh5v3i-story.html** (Accessed August 3, 2020).

353 Ceruzzi, P. E. (2003). A History of Modern Computing (W. Aspray, Ed.; second edition). The MIT Press.

354 Waldrop, M. (2008). DARPA and the Internet revolution. DARPA: 50 Years of Bridging the Gap, (December 1969), 78–85. **https://www.darpa.mil/attachments/(2O15)%20Global%20Nav%20-%20About%20Us%20-%20History%20-%20Resources%20-%2050th%20-%20Internet%20(Approved).pdf**

355 Sputnik 1. (2020). Wikipedia. **https://en.wikipedia.org/w/index.php?title=Sputnik_1&oldid=960627959** (Accessed June 12, 2020).

356 Rewire Security Team. (2019). Origin of Global Positioning System (GPS) Rewire Security. Rewire Security. **https://www.rewiresecurity.co.uk/blog/gps-global-positioning-system-satellites** (Accessed June 12, 2020).

357 Platt, J. R. (1964). Strong Inference: Certain systematic methods of scientific thinking may produce much more rapid progress than others. Science, 146(3642), 347–353. **https://doi.org/10.1126/science.146.3642.347.**

Judson, H. F. (1996). The Eighth Day of Creation: Makers of the Revolution in Biology, Commemorative Edition (Expanded edition). Cold Spring Harbor Laboratory Press

358 Direct DNA Damage. (2020). Wikipedia. **https://en.wikipedia.org/w/index.php?title=Direct_DNA_damage&oldid=944315847** (Accessed August 3, 2020).

359 Arrow, K. (1962). Economic Welfare and the Allocation of Resources for Invention. In The Rate and Direction of Inventive Activity: Economic and Social Factors (pp. 609–626).

360 Truong, A. (2015). Huawei's R&D Spend Is Massive Even by the Standards of American Tech Giants. Quartz. **https://qz.com/374039/huaweis-rd-spend-is-massive-even-by-the-standards-of-american-tech-giants/** (Accessed June 12, 2020).

361 What Percent of Revenue Do Publicly Traded Companies Spend on Marketing and Sales? (2020). Vital Design. **https://vtldesign.com/digital-marketing/content-marketing-strategy/percent-of-revenue-spent-on-marketing-sales/** (Accessed June 12, 2020).

362 Rooney, K. (2019). Share Buybacks Soar to Record $806 Billion — Bigger than a Facebook or Exxon Mobil. CNBC. **https://www.cnbc.com/2019/03/25/share-buybacks-soar-to-a-record-topping-800-billion-bigger-than-a-facebook-or-exxon-mobil.html** (Accessed June 12, 2020).

Global revenue of Apple from 2004 to 2020. Statista. **https://www.statista.com/statistics/265125/total-net-sales-of-apple-since-2004/** (Accessed June 12, 2020).

363 Mazzucato, M. (2015). The Entrepreneurial State: Debunking Public vs. Private Sector Myths (Revised edition). PublicAffairs.

364 Horan, H. (2019). Uber's Path of Destruction. American Affairs Journal. **https://americanaffairsjournal.org/2019/05/ubers-path-of-destruction/** (Accessed November 27, 2020).

Eavis, P. (2019). 'It's Definitely Pretty Empty': Why Saving WeWork Will Be Hard. The New York Times. **https://www.nytimes.com/2019/10/24/business/wework-growth.html** (Accessed November 27, 2020).

365 TESLA. Energy.gov. **https://www.energy.gov/lpo/tesla** (Accessed June 13, 2020).

366 Tesla Market Cap 2009-2021 | TSLA. Macrotrends. **https://www.macrotrends.net/stocks/charts/TSLA/tesla/market-cap** (Accessed June 13, 2020).

367 Howell, K., & Dinen, S. (2015, August 26). Solyndra misled government to get $535M solar project loan. The Washington Times. **https://www.washingtontimes.com/news/2015/aug/26/solyndra-misled-government-get-535-million-solar-p/** (Accessed June 13, 2020).

368 Brady, J. (2014, November 13). After Solyndra Loss, U.S. Energy Loan Program Turning A Profit. NPR.Org. **https://www.npr.org/2014/11/13/363572151/after-solyndra-loss-u-s-energy-loan-program-turning-a-profit** (Accessed June 13, 2020).

369 Graham, P. (2013). How to Convince Investors. **http://www.paulgraham.com/convince.html** (Accessed June 13, 2020).

370 Ford, G. S., Koutsky, T., & Spiwak, L. J. (2007). A Valley of Death in the Innovation Sequence: An Economic Investigation (SSRN Scholarly Paper ID 1093006). Social Science Research Network. **https://doi.org/10.2139/ssrn.1093006.**

371 Press, L. (2012). Government Spending: Seeding the Internet Cost $124.5 Million, Morse's Telegraph $30 Thousand. **http://cis471.blogspot.com/2012/08/seeding-internet-cost-government-1245.html** (Accessed August 3, 2020).

372 Warner, M. (2019). Industrial Biotechnology Commercialization Handbook: How to make proteins without animals and fuels or chemicals without crude oil. Independently published.

373 Warner, M. (2020). Successfully commercializing alternative proteins - bench to plate with Mark Warner. Good Food Institute. Presented September 18th, 2020. **https://www.youtube.com/watch?v=wJcwEXatj04**

374 Wiebe, M. (2004). Quorn™ Myco-protein — Overview of a successful fungal product. Mycologist, 18(1), 17–20. **https://doi.org/10.1017/S0269-915X(04)00108-9.**

375 Our Impact. ARPA-E. **https://arpa-e.energy.gov/?q=site-page/arpa-e-impact** (Accessed June 13, 2020).

376 Roberts, D. (2019). Jay Inslee Is Writing the Climate Plan the next President Should Adopt. Vox. **https://www.vox.com/energy-and-environment/2019/5/18/18628870/green-new-deal-jay-inslee-2020-climate-change** (Accessed June 13, 2020).

377 Clancy, S. (2008) Genetic mutation. Nature Education 1(1):187 **https://www.nature.com/scitable/topicpage/genetic-mutation-441/**

378 Deltcheva, E., Chylinski, K., Sharma, C. M., Gonzales, K., Chao, Y., Pirzada, Z. A., … Charpentier, E. (2011). CRISPR RNA maturation by trans-encoded small RNA and host factor RNase III. Nature, 471(7340), 602–607. **https://doi.org/10.1038/nature09886.**

379 Jacob, F. (1995). The Statue Within: An Autobiography (Illustrated edition). Cold Spring Harbor Laboratory Press.

380 Yanai, I., & Lercher, M. (2019). Night science. Genome Biology, 20(1), 20–22. **https://doi.org/10.1186/s13059-019-1800-6.**

381 Suckjoon Jun. University of California – San Diego. **https://biology.ucsd.edu/research/faculty/s2jun** (Accessed January 13, 2021).

Suckjoon Jun also coauthored a provoking essay explaining his philosophy on night science and scientific funding:

Desai, A., & Jun, S. (2018). Promoting an "Auteur Theory" for Young Scientists: Preserving Excitement and Creativity …. BioEssays, 40(11), 1800147. **https://doi.org/10.1002/bies.201800147.**

382 Taheri-Araghi, S., Bradde, S., Sauls, J. T., Hill, N. S., Levin, P. A., Paulsson, J., … Jun, S. (2015). Cell-size control and homeostasis in bacteria. Current Biology, 25(3), 385–391. **https://doi.org/10.1016/j.cub.2014.12.009.**

Si, F., Li, D., Cox, S. E., Sauls, J. T., Azizi, O., Sou, C., … Jun, S. (2017). Invariance of Initiation Mass and Predictability of Cell Size in Escherichia coli. Current Biology, 27(9), 1278–1287. **https://doi.org/10.1016/j.cub.2017.03.022.**

383 Wang, P., Robert, L., Pelletier, J., Dang, W. L., Taddei, F., Wright, A., & Jun, S. (2010). Robust growth of Escherichia coli. Current Biology : CB, 20(12), 1099–1103. **https://doi.org/10.1016/j.cub.2010.04.045.**

384 Jun, S. Jun Lab | qBio | UCSD Physics and Molecular Biology. **https://jun.ucsd.edu/** (Accessed January 13, 2021).

385 Sekar, K., Rusconi, R., Sauls, J. T., Fuhrer, T., Noor, E., Nguyen, J., … Sauer, U. (2018). Synthesis and degradation of FtsZ quantitatively predict the first cell division in starved bacteria. Molecular Systems Biology, 14(11), 8623. **https://doi.org/10.15252/msb.20188623.**

Si, F., Le Treut, G., Sauls, J. T., Vadia, S., Levin, P. A., & Jun, S. (2019). Mechanistic Origin of Cell-Size Control and Homeostasis in Bacteria. Current Biology, 29(11), 1760-1770.e7. **https://doi.org/10.1016/j.cub.2019.04.062.**

386 ETH Zurich Research Commission. **https://ethz.ch/en/the-eth-zurich/organisation/boards-university-groups-commissions/research-commission.html** (Accessed January 13, 2021).

387 Pink, D. (2009). The puzzle of motivation. TED – ideas worth spreading. **https://www.ted.com/talks/dan_pink_the_puzzle_of_motivation** (Accessed June 13, 2020).

388 Hourihan, M., & Parkes, D. (2016). Federal R & D Budget Trends : A Short Summary. (January), 1–7.

389 Pepino, M. Y., Love-Gregory, L., Klein, S., & Abumrad, N. A. (2012). The fatty acid translocase gene CD36 and lingual lipase influence oral sensitivity to fat in obese subjects. Journal of Lipid Research, 53(3), 561–566. **https://doi.org/10.1194/jlr.M021873.**

390 Bockisch, M. (Ed.). (1998). Chapter 3—Animal Fats and Oils. In Fats and Oils Handbook (pp. 121–173). AOCS Press. **https://doi.org/10.1016/B978-0-9818936-0-0.50008-1.**

391 List, G., & Jackson, M. (2007). The Battle Over Centralization (1903-1920). INFORM - International News on Fats, Oils and Related Materials., 18, 403–405. **https://doi.org/10.1016/j.tej.2012.03.006.**

392 Eun, S. J., Mun, Y. J., & Min, D. B. (2005). Hydrogenation for low trans and high conjugated fatty acids. Comprehensive Reviews in Food Science and Food Safety, Vol. 4, pp. 22–30. **https://doi.org/10.1111/j.1541-4337.2005.tb00069.x.**

393 Trumbo, P., Schlicker, S., Yates, A. A., Poos, M., & Food and Nutrition Board of the Institute of Medicine, The National Academies. (2002). Dietary reference intakes for energy, carbohydrate, fiber, fat, fatty acids, cholesterol, protein and amino acids. Journal of the American Dietetic Association, 102(11), 1621–1630. **https://doi.org/10.1016/s0002-8223(02)90346-9.**

394 Removing Trans Fats from McDonald's® Famous Fries. (2015). Cargill. **https://www.cargill.com/history-story/en/HIGH-OLEIC.jsp** (June 13, 2020).

395 Aznar-Moreno, J. A., & Durrett, T. P. (2017). Review: Metabolic engineering of unusual lipids in the synthetic biology era. Plant Science: An International Journal of Experimental Plant Biology, 263, 126–131. **https://doi.org/10.1016/j.plantsci.2017.07.007.**

396 Vanhercke, T., Wood, C. C., Stymne, S., Singh, S. P., & Green, A. G. (2013). Metabolic engineering of plant oils and waxes for use as industrial feedstocks. Plant Biotechnology Journal, Vol. 11, pp. 197–210. **https://doi.org/10.1111/pbi.12023.**

Soong, Y. H. V., Liu, N., Yoon, S., Lawton, C., & Xie, D. (2019). Cellular and metabolic engineering of oleaginous yeast Yarrowia lipolytica for bioconversion of hydrophobic substrates into high-value products. Engineering in Life Sciences, Vol. 19, pp. 423–443. **https://doi.org/10.1002/elsc.201800147.**

397 Omega-3 Fatty Acids. (2020). National Institute of Health - Office of Dietary Supplements. **https://ods.od.nih.gov/factsheets/Omega3FattyAcids-HealthProfessional/** (Accessed August 4, 2020).

398 National Research Council (US) Committee on Diet and Health. (1989). Fat-Soluble Vitamins. In Diet and Health: Implications for Reducing Chronic Disease Risk. National Academies Press (US). **https://www.ncbi.nlm.nih.gov/books/NBK218749/.**

399 Nicholson, R. A., & Marangoni, A. G. (2020). Enzymatic glycerolysis converts vegetable oils into structural fats with the potential to replace palm oil in food products. Nature Food, 1(November). **https://doi.org/10.1038/s43016-020-00160-1.**

400 Raza, S., Fransson, L., & Hult, K. (2001). Enantioselectivity in Candida antarctica lipase B: A molecular dynamics study. Protein Science: A Publication of the Protein Society, 10(2), 329–338. **https://doi.org/10.1110/ps.33901.**

401 Flickinger, B. D., & Matsuo, N. (2003). Nutritional characteristics of DAG oil. Lipids, 38(2), 129–132. **https://doi.org/10.1007/s11745-003-1042-8.**

402 Ah, J., & Tagalpallewar, G. P. (2017). Functional properties of Mozzarella cheese for its end use application. Journal of Food Science and Technology, 54(12), 3766–3778. **https://doi.org/10.1007/s13197-017-2886-z.**

403 How We Make Animal-Free Dairy Proteins. Perfect Day. **https://www.perfectdayfoods.com/how-it-works/** (Accessed June 14, 2020).

New Culture. New Culture. **https://www.newculturefood.com/** (Accessed June 14, 2020).

404 Hristov, P., Mitkov, I., Sirakova, D., Mehandgiiski, I., & Radoslavov, G. (2016). Measurement of Casein Micelle Size in Raw Dairy Cattle Milk by Dynamic Light Scattering. Milk Proteins - From Structure to Biological Properties and Health Aspects, (September). **https://doi.org/10.5772/62779.**

405 Pierce, M. M., Raman, C. S., & Nall, B. T. (1999). Isothermal titration calorimetry of protein-protein interactions. Methods (San Diego, Calif.), 19(2), 213–221. **https://doi.org/10.1006/meth.1999.0852.**

Label Free BLI Detection. ForteBio. **https://www.fortebio.com/products/label-free-bli-detection** (Accessed June 14, 2020).

Kanai, T., Egoshi, K., Ohno, S., & Takebe, T. (2018). The evaluation of stretchability and its applications for biaxially oriented polypropylene film. Advances in Polymer Technology, 37(6), 2253–2260. **https://doi.org/10.1002/adv.21884.**

Fife, R. L., McMahon, D. J., & Oberg, C. J. (2002). Test for measuring the stretchability of melted cheese. Journal of Dairy Science, 85(12), 3539–3545. **https://doi.org/10.3168/jds.S0022-0302(02)74444-5.**

406 Marti, D., Johnson, R., & Mathews, K. (2011, November). Where's the (Not) Meat?-Byproducts From Beef and Pork Production. USDA Economic Research Service. **http://www.ers.usda.gov/publications/pub-details/?pubid=37428** (Accessed August 4, 2020).

407 Tubb, C., & Seba, T. (2019). Rethinking Food and Agriculture 2020-2030. RethinkX. **https://www.rethinkx.com/food-and-agriculture.**

408 Abd El-Hady, R.A.M & Abd El-Baky, R.A.A. (2011). Enhancing the Functional Properties of Sportswear Fabric based Carbon Fiber. Asian Journal of Textile. 1(1), 14-26. **https://doi.org/10.3923/ajt.2011.14.26.**

409 Mulvany, L. & Rupp, L. (2018). As Leather Shoes Drop out of Favor, Cattle Hides Pile Up. Los Angeles Times. **https://www.latimes.com/business/la-fi-leather-shoes-20180612-story.html** (Accessed August 4, 2020).

410 Koran, M. (2019). Macy's Becomes Biggest US Retailer to End Fur Sales. The Guardian. **http://www.theguardian.com/business/2019/oct/21/macys-ends-real-fur-sales** (Accessed August 4, 2020).

411 Dallmeier, L. (2013). 13 Animal Products in Cosmetics. Herb & Hedgerow. **http://www.herbhedgerow.co.uk/animal-products-in-cosmetics/** (Accessed August 4, 2020).

412 Newton, A. A. (2019). Do Collagen Creams and Supplements Actually Do Anything? SELF. **https://www.self.com/story/collagen-creams-supplements-skin** (Accessed June 14, 2020).

413 Kirkova, D. (2013). Lily Cole to Reveal the Ugly Truth behind Luxury Beauty: Model Exposes Cosmetics Industry's Cruel Use of SHARK Liver. Mail Online. **https://www.dailymail.co.uk/femail/article-2331563/Lily-Cole-reveal-ugly-truth-luxury-beauty-Model-exposes-cosmetics-industrys-cruel-use-SHARK-liver.html** (Accessed August 4, 2020).

414 Li, T., Liu, G.-S., Zhou, W., Jiang, M., Ren, Y.-H., Tao, X.-Y., Liu, M., Zhao, M., Wang, F.-Q., Gao, B., & Wei, D.-Z. (2020). Metabolic Engineering of Saccharomyces cerevisiae To Overproduce Squalene. Journal of Agricultural and Food Chemistry, 68(7), 2132–2138. **https://doi.org/10.1021/acs.jafc.9b07419**.

415 Home | Geltor | Biodesigned solutions for beauty, nutrition, food & beverage. Geltor. **https://geltor.com/** (Accessed June 14, 2020).

416 Ro, D. K., Paradise, E. M., Quellet, M., Fisher, K. J., Newman, K. L., Ndungu, J. M., ... Keasling, J. D. (2006). Production of the antimalarial drug precursor artemisinic acid in engineered yeast. Nature, 440(7086), 940–943. **https://doi.org/10.1038/nature04640**.

CHAPTER 10

417 Eveleth, R. (2012). What Will Convince People That Genetically Modified Foods Are Okay? Smithsonian Magazine. **https://www.smithsonianmag.com/smart-news/what-will-convince-people-that-genetically-modified-foods-are-okay-125478012/** (August 5, 2020).

418 Kolodinsky, J., & Lusk, J. L. (2018). Mandatory labels can improve attitudes toward genetically engineered food. Science Advances, 4(6), 1–6. **https://doi.org/10.1126/sciadv.aaq1413**.

419 Foer, J. S. (2010). Eating Animals (1st edition). Back Bay Books.

420 What Is Ag-Gag Legislation? ASPCA. **https://www.aspca.org/animal-protection/public-policy/what-ag-gag-legislation** (Accessed July 4, 2020).

421 Bollard, L. (2013). The Terrible Price of Ag-Gag Laws. New York Daily News. **https://www.nydailynews.com/opinion/terrible-price-ag-gag-laws-article-1.1346292** (Accessed July 4, 2020).

422 Hall, C. (2013). California 'ag-Gag Bill' Is Gagged and Gone. Los Angeles Times. **https://www.latimes.com/opinion/la-xpm-2013-apr-17-la-ol-california-aggag-bill-is-gagged-and-gone-20130417-story.html** (Accessed July 4, 2020).

Doran, W. (2020). In Targeting Animal Rights Activists, NC Violated the First Amendment, Court Rules. The News & Observer. **https://www.newsobserver.com/news/politics-government/article243547507.html** (Accessed July 4, 2020).

423 EC Council Directive. (1999). Council Directive 99/74/EC of 19 July 1999 laying down minimum standards for the protection of laying hens. Official Journal of the European Communities, (6), 53–57. **http://data.europa.eu/eli/dir/1999/74/oj.**

424 Sherwin, C. M., Richards, G. J., & Nicol, C. J. (2010). Comparison of the welfare of layer hens in 4 housing systems in the UK. British Poultry Science, 51(4), 488–499. **https://doi.org/10.1080/00071668.2010.502518.**

425 Grandin, T. (2015). Improving Animal Welfare: A Practical Approach (2nd edition). CABI.

Farm Animal Confinement Bans by State. ASPCA. **https://www.aspca.org/animal-protection/public-policy/farm-animal-confinement-bans** (Accessed July 4, 2020).

426 Little, J.B. (2009). The Ogallala Aquifer: Saving a Vital U.S. Water Source. Scientific American. **https://www.scientificamerican.com/article/the-ogallala-aquifer/** (Accessed January 14, 2021).

427 Amadeo, K. (2020). The Dust Bowl, Its Causes, Impact, With a Timeline and Map. The Balance. **https://www.thebalance.com/what-was-the-dust-bowl-causes-and-effects-3305689** (Accessed July 5, 2020).

Droughts in the United States. (2020). Wikipedia. **https://en.wikipedia.org/w/index.php?title=Droughts_in_the_United_States&oldid=959172389** (Accessed July 5, 2020).

428 Growing A Nation. **https://growinganation.org/** (July 5, 2020).

Trimarchi, M. (2008). What Caused the Dust Bowl?. HowStuffWorks. **https://science.howstuffworks.com/environmental/green-science/dust-bowl-cause.htm** (Accessed July 5, 2020).

429 Imhoff, D., & Badaracco, C. (2019). The Farm Bill: A Citizen's Guide (3rd edition). Island Press.

430 The Farm Bill is not a singular piece of legislation, but rather deceptively refers to a series of legislation over the years that affect agriculture policy (e.g., Food and Agriculture Act of 1977, Federal Agriculture Improvement and Reform Act of 1996, etc.).

431 Hoppe, R. A. (2015). Structure and finances of U.S. farms: Family farm report, 2014 edition. U.S. Family Farms: Structure, Finances, and Agricultural Production Role, (132), 1–91.

432 Bakst, D., & Wright, B. (2016). Addressing Risk in Agriculture. The Heritage Foundation, (189), 64.

433 Imhoff, D., & Badaracco, C. (2019). The Farm Bill: A Citizen's Guide (3rd edition). Island Press.

434 Starmer, E., & Wise, T. A. (2007). Feeding at the Trough: Industrial Livestock Firms Saved $35 billion From Low Feed Prices. In GDAE Policy Brief (Vol. 7).

435 Database, EWG's Farm Subsidy. EWG's Farm Subsidy Database. **http://farm.ewg.org/progdetail.php?fips=00000&progcode=corn** (Accessed July 5, 2020).

Honig, L. (2015). August Crop Production Executive Summary. In National Agricultural Statistics Service.

436 Soybean Subsidies in the United States Totaled $46.1 Billion from 1995-2019. EWG's Farm Subsidy Database. **https://farm.ewg.org/progdetail.php?fips=00000&progcode=soybean** (Accessed July 5, 2020).

437 Wills, K. (2013). Where Do All These Soybeans Go? MSU Extension. **https://www.canr.msu.edu/news/where_do_all_these_soybeans_go** (Accessed July 5, 2020).

438 Margin Protection Program for Dairy; MPP Dairy. USDA – Farm Service Agency. **https://www.fsa.usda.gov/programs-and-services/Dairy-MPP/index** (Accessed July 5, 2020).

439 What Happens to Animal Waste? FoodPrint. **https://foodprint.org/issues/what-happens-to-animal-waste/** (Accessed July 5, 2020).

440 Environmental Quality Incentives Program (EQIP) | Farm Bill Report (FY 2009 through FY 2019). USDA – Natural Resources Conservation Service. **https://www.nrcs.usda.gov/Internet/NRCS_RCA/reports/fb08_cp_eqip.html** (Accessed July 5, 2020).

441 EQIP: The Farm Bill, Livestock Producers & Help to Pay For Your Building. Summit Livestock Facilities. **https://www.summitlivestock.com/news-events/eqip-the-farm-bill-livestock-producers-help-to-pay-for-your-building/** (Accessed July 5, 2020).

442 Yeh, C. Y., Schafferer, C., Lee, J. M., Ho, L. M., & Hsieh, C. J. (2017). The effects of a rise in cigarette price on cigarette consumption, tobacco taxation revenues, and of smoking-related deaths in 28 EU countries - Applying threshold regression modelling. BMC Public Health, 17(1), 1–9. **https://doi.org/10.1186/s12889-017-4685-x.**

Xu, X., & Chaloupka, F. J. (2011). The effects of prices on alcohol use and its consequences. Alcohol Research & Health : The Journal of the National Institute on Alcohol Abuse and Alcoholism, 34(2), 236–245. **http://www.ncbi.nlm.nih.gov/pubmed/22330223.**

Knittel, C. R., & Tanaka, S. (2019). Driving Behaviour and the Price of Gasoline: Evidence from Fueling-Level Micro Data. NBER Working Paper, November, 1–19.

443 Bakst, D., & Wright, B. (2016). Addressing Risk in Agriculture. The Heritage Foundation, (189), 64.

444 Imhoff, D., & Badaracco, C. (2019). The Farm Bill: A Citizen's Guide (3rd edition). Island Press.

445 Nordhaus, W. D. (1996). The Economics of New Goods. In T. F. Bresnahan & R. J. Gordon (Eds.), The Economics of New Goods (pp. 27–70). **http://www.nber.org/books/bres96-1**.

446 Common Agricultural Policy. (2020). Wikipedia. **https://en.wikipedia.org/w/index.php?title=Common_Agricultural_Policy&oldid=964054243** (Accessed August 5, 2020).

447 Amadeo, K. (2020). US Imports and Exports with Components and Statistics. What Does the United States Trade With Foreign Countries? The Balance. **https://www.thebalance.com/u-s-imports-and-exports-components-and-statistics-3306270** (Accessed July 5, 2020).

448 Gilbert, B. (2019). The Impossible burger has 1 major flaw to overcome: It's nearly triple the price of normal ground beef. Business Insider. **https://www.businessinsider.com/the-impossible-burger-is-too-expensive-2019-9?op=1** (Accessed August 29, 2020).

Kirkwood, B. (2020). Beyond Meat Beyond Burgers Drop Price To $1.60 A Burger. Vegan News. **https://vegannewsnow.com/2020/06/19/beyond-meat-price-drop/** (Accessed August 29, 2020).

449 Reese, J. (2018). The End of Animal Farming: How Scientists, Entrepreneurs, and Activists Are Building an Animal-Free Food System. Beacon Press.

450 Ferguson, D. (2019). Vegan College Menus on the Rise as Students Return to University. The Guardian. **http://www.theguardian.com/lifeandstyle/2019/sep/22/vegan-college-menus-on-rise-as-students-return-to-universitys** (Accessed July 11, 2020).

451 Wiltsie, M. (2018). Report: Two-Thirds of Colleges Now Offer Three Square Vegan Meals a Day. PETA. **https://www.peta.org/media/news-releases/report-two-thirds-of-colleges-now-offer-three-square-vegan-meals-a-day/** (Accessed July 11, 2020).

452 Check out The University of North Carolina at Chapel Hill's Vegan Report Card Grade! PETA's Vegan Report Card. **https://collegereportcard.peta.org/college/the-university-of-north-carolina-at-chapel-hill/** (Accessed July 11, 2020).

453 Butler, K. (2015). The Surprising Reason Why School Cafeterias Sell Chocolate Milk. Mother Jones. **https://www.motherjones.com/environment/2015/11/milk-companies-market-schools-fast-food/** (Accessed July 11, 2020).

454 Purdy, C. (2020, April 6). Covid-19 has the US cheese and milk industries on the brink. Quartz. **https://qz.com/1832063/covid-19-has-us-cheese-and-milk-industries-on-the-brink/** (Accessed July 11, 2020).

455 Nutrition Standards for School Meals. USDA – Food and Nutrition Service. **https://www.fns.usda.gov/school-meals/nutrition-standards-school-meals** (Accessed July 11, 2020).

456 National School Lunch Program (NSLP) Fact Sheet. USDA – Food and Nutrition Service. **https://www.fns.usda.gov/nslp/nslp-fact-sheet** (Accessed July 11, 2020).

457 News 12 Staff. (2020). Brooklyn Borough President Proposes Non-Dairy Milk Choices for School Children. News 12 Brooklyn. **http://brooklyn.news12.com/story/41854679/brooklyn-borough-president-proposes-nondairy-milk-choices-for-school-children** (Accessed July 11, 2020).

458 USDA Working with Dairy Industry to Ensure Americans' Consistent Access to Milk. USDA - Agricultural Marketing Service. **https://www.ams.usda.gov/content/usda-working-dairy-industry-ensure-americans%E2%80%99-consistent-access-milk** (Accessed July 11, 2020).

459 Melnick, M. (2011). The USDA Ditches the Food Pyramid for a Plate. TIME. **https://healthland.time.com/2011/06/02/the-usda-ditches-the-food-pyramid-and-offers-a-plate/** (Accessed July 11, 2020).

460 Price Support. USDA – Farm Service Agency. **https://fsa.usda.gov/programs-and-services/price-support/Index** (Accessed January 15, 2021).

461 The Reducetarian Foundation **https://www.reducetarian.org** (Accessed July 12, 2020).

462 Singer, P. (2015). Animal Liberation: The Definitive Classic of the Animal Movement. Open Road Media.

463 Lane. (2020). How Many Vegans in The World? In the USA? (2020). Vegan Bits. **https://veganbits.com/vegan-demographics/** (Accessed July 12, 2020).

464 Parker, J. (2019). The Year of the Vegan. Economist. **https://www.bluehorizon.com/economist-names-2019-the-year-of-vegan/** (Accessed July 12, 2020).

465 ModVegan. (2017). The S-Curve of Innovation & the End of Animal Agriculture. ModVegan. **https://modvegan.com/s-curve/** (Accessed July 12, 2020).

466 Schaeffer, K. (2020). Key Facts about Women's Suffrage around the World, a Century after U.S. Ratified 19th Amendment. Pew Research Center. **https://www.pewresearch.org/fact-tank/2020/10/05/key-facts-about-womens-suffrage-around-the-world-a-century-after-u-s-ratified-19th-amendment/** (Accessed January 15, 2021).

History.com Editors. (2021). Civil Rights Movement. HISTORY. **https://www.history.com/topics/black-history/civil-rights-movement** (Accessed January 15, 2021).

467 Higson, A. (2019). Vegan Stigma: A Barrier To Dietary Change. Faunalytics. **https://faunalytics.org/vegan-stigma-a-barrier-to-dietary-change/** (Accessed July 12, 2020).

468 Choi, C. (2018). 'Plant-Based' Replaces 'v-Words' to Appeal to Carnivores. Christian Science Monitor. **https://www.csmonitor.com/The-Culture/Food/2018/0824/Plant-based-replaces-v-words-to-appeal-to-carnivores** (Accessed July 12, 2020).

469 King, B. (2012). Do Vegetarians And Vegans Think They Are Better Than Everyone Else? NPR.org. **https://www.npr.org/sections/13.7/2012/08/30/160117028/do-vegetarians-and-vegans-think-they-are-better-than-everyone-else** (Accessed July 12, 2020).

470 Bricker, K. (2017). How Can You Tell If Someone Is Vegan? Don't Worry, They'll Tell You. Medium. **https://medium.com/@korinnebricker/how-can-you-tell-if-someone-is-vegan-dont-worry-they-ll-tell-you-4eec13254f03** (Accessed July 12, 2020).

471 Jasiunas, L. (2020). Veganism, Stigma, And You. Faunalytics. **https://faunalytics.org/veganism-stigma-and-you/** (Accessed August 29, 2020).

472 Hidden Brain: America's Changing Attitudes Toward Gay People. NPR.org. **https://www.npr.org/2019/04/17/714212984/hidden-brain-americas-changing-attitudes-toward-gay-people** (Accessed July 12, 2020).

473 Ettinger, J. (2019). 50 Celebrities Who Are Vegan For Life. LIVEKINDLY. **https://www.livekindly.co/vegan-celebrities/** (Accessed July 12, 2020).

474 Castrodale, J. (2019). Vegan Bride Bans All Non-Vegans from Her Wedding for Being 'Murderers.' **https://www.vice.com/en_us/article/7xnwyq/vegan-bride-bans-all-non-vegans-from-her-wedding-for-being-murderers** (Accessed July 12, 2020).

475 r/Insanepeoplefacebook - This Lady Banned All Non-Vegans from Her Wedding, Including Family and Bridal Party. Reddit. **https://www.reddit.com/r/insanepeoplefacebook/comments/andl11/this_lady_banned_all_nonvegans_from_her_wedding/** (Accessed July 12, 2020).

476 Brion, R. (2011). David Chang on Treme: 'Let's Put Pork in Every Fucking Dish.' Eater. **https://www.eater.com/2011/6/20/6673991/david-chang-on-treme-lets-put-pork-in-every-fucking-dish** (Accessed January 15, 2021).

477 r/Vegan - What Are Your Thoughts about the Vegan Bride That Banned Her Meat-Eating Family Members from Attending Her Wedding? Reddit. **https://www.reddit.com/r/vegan/comments/ajeps5/what_are_your_thoughts_about_the_vegan_bride_that/** (Accessed July 12, 2020).

478 Knowing Gays and Lesbians, Religious Conflicts, Beliefs about Homosexuality. (2015). Pew Research Center - U.S. Politics & Policy. **https://www.pewresearch.org/politics/2015/06/08/section-2-knowing-gays-and-lesbians-religious-conflicts-beliefs-about-homosexuality/** (Accessed July 12, 2020).

Feelings toward religious groups. (2019). Pew Research Center's Religion & Public Life Project. **https://www.pewforum.org/2019/07/23/feelings-toward-religious-groups/** (Accessed July 12, 2020).

CHAPTER 11

479 Whale Whores. (2020). Wikipedia. **https://en.wikipedia.org/w/index.php?title=Whale_Whores&oldid=958406814** (Accessed July 18, 2020).

480 Liao, M. S. (2011). Bias and Reasoning: Haidt's Theory of Moral Judgment. In New Waves in Ethics (2011 edition). Palgrave Macmillan.

481 Kurzgesagt – In a Nutshell. (2018, September 30). Why Meat is the Best Worst Thing in the World. **https://www.youtube.com/watch?v=NxvQPzrg2Wg** (Accessed July 12, 2020).

482 Kurzgesagt – In a Nutshell. (2020, January 26). Milk. White Poison or Healthy Drink? **https://www.youtube.com/watch?v=oakWgLqCwUc** (Accessed July 12, 2020).

483 Joint Statement. (2001). The Science of Climate Change. Science, 292(5520), 1261–1261. **https://doi.org/10.1126/science.292.5520.1261**

A case of junk science, conflict and hype. (2008). Nature Immunology, 9(12), 1317–1317. **https://doi.org/10.1038/ni1208-1317.**

484 Chen, S. (2020). Researchers around the World Prepare to #ShutDownSTEM and 'Strike For Black Lives.' Science | AAAS. **https://www.sciencemag.org/news/2020/06/researchers-around-world-prepare-shutdownstem-and-strike-black-lives** (Accessed August 6, 2020).

Wadman, M. (2018). Scientists Share MIT 'Disobedience' Award for #MeToo Advocacy. Science | AAAS. **https://www.sciencemag.org/news/2018/11/metoo-advocates-share-mit-disobedience-award** (Accessed August 6, 2020).

485 Singer, P. (2015). Animal Liberation: The Definitive Classic of the Animal Movement. Open Road Media.

486 Vinding, M. (2014). Why We Should Go Vegan.

487 Moral Relativism. (2020). Wikipedia. **https://en.wikipedia.org/w/index.php?title=Moral_relativism&oldid=970024964** (Accessed August 29, 2020).

488 Sati (Practice). (2020). Wikipedia. **https://en.wikipedia.org/w/index.php?title=Sati_(practice)&oldid=971480686** (Accessed August 6, 2020).

Riales, D. (2017). Modern Day Witch Hunt Kills Hundreds of Elderly African Women. Blasting News. **https://us.blastingnews.com/curiosities/2017/08/modern-day-witch-hunt-kills-hundreds-of-elderly-african-women-001898617.html** (Accessed August 6, 2020).

489 Human Rights in Qatar. (2020). Wikipedia. **https://en.wikipedia.org/w/index.php?title=Human_rights_in_Qatar&oldid=970142944** (Accessed August 6, 2020).

490 Peat, Jack. (2017). Government Votes That 'Animals Can't Feel Pain or Emotions.' The London Economic. **https://www.thelondoneconomic.com/news/government-votes-animals-cant-feel-pain-emotions/17/11/** (Accessed July 12, 2020).

491 Pain in Animals. (2020). Wikipedia. **https://en.wikipedia.org/w/index.php?title=Pain_in_animals&oldid=960958456** (Accessed July 12, 2020).

492 Allen, C. (1998). Assessing animal cognition: Ethological and philosophical perspectives. Journal of Animal Science, 76(1), 42–47. **https://doi.org/10.2527/1998.76142x**

Carbone, L. (2004). What Animals Want: Expertise and Advocacy in Laboratory Animal Welfare Policy (1st edition). Oxford University Press.

493 Hard Problem of Consciousness. (2020). Wikipedia. **https://en.wikipedia.org/w/index.php?title=Hard_problem_of_consciousness&oldid=964471782** (Accessed August 6, 2020).

494 Singer, P. (2015). Animal Liberation: The Definitive Classic of the Animal Movement. Open Road Media.

495 Conklin, B. A. (2001). Consuming Grief: Compassionate Cannibalism in an Amazonian Society. University of Texas Press.

496 Associated Press. (2008). China Protesters: Stop 'Cooking Cats Alive.' MSNBC News. **http://www.nbcnews.com/id/28292558/ns/world_news-asia_pacific/t/china-protesters-stop-cooking-cats-alive/** (Accessed July 18, 2020).

Cochrane, J. (2017). Indonesians' Taste for Dog Meat Is Growing, Even as Others Shun It. The New York Times. **https://www.nytimes.com/2017/03/25/world/asia/indonesia-dog-meat.html** (Accessed July 18, 2020).

497 Hinckley, S. (2016). How Yao Ming Appeased the Chinese Appetite for Shark Fin Soup. Christian Science Monitor. **https://www.csmonitor.com/World/Global-News/2016/0602/How-Yao-Ming-appeased-the-Chinese-appetite-for-shark-fin-soup** (Accessed July 18, 2020).

498 Day, D. (1987). The Whale War. First Edition. San Francisco: Random House, Inc.

499 Liao, M. S. (2011). Bias and Reasoning: Haidt's Theory of Moral Judgment. In New Waves in Ethics (2011 edition). Palgrave Macmillan.

500 Species. (2020). Wikipedia. **https://en.wikipedia.org/w/index.php?title=Species&oldid=971131740** (Accessed August 6, 2020).

501 Church, G. M., & Regis, E. (2014). Regenesis: How Synthetic Biology Will Reinvent Nature and Ourselves (1st edition). Basic Books.

502 Qualman Darrin. (2018). Earth's Dominant Bird: A Look at 100 Years of Chicken Production. **https://www.darrinqualman.com/100-years-chicken-production/** (Accessed July 19, 2020).

503 Veal. (2020). Wikipedia. **https://en.wikipedia.org/w/index.php?title=Veal&oldid=962432952** (Accessed August 6, 2020).

504 Anthis, J. R. (2019). US Factory Farming Estimates. Sentience Institute. **https://sentienceinstitute.org/us-factory-farming-estimates** (Accessed September 20, 2020).

505 Creel, S., Christianson, D., Liley, S., & Winnie, J. A. (2007). Predation Risk Affects Reproductive Physiology and Demography of Elk. Science, 315(5814), 960–960. **https://doi.org/10.1126/science.1135918.**

506 Creel, S., Winnie, J. A., & Christianson, D. (2009). Glucocorticoid stress hormones and the effect of predation risk on elk reproduction. Proceedings of the National Academy of Sciences, 106(30), 12388–12393. **https://doi.org/10.1073/pnas.0902235106.**

507 Zanette, L. Y., Hobbs, E. C., Witterick, L. E., MacDougall-Shackleton, S. A., & Clinchy, M. (2019). Predator-induced fear causes PTSD-like changes in the brains and behaviour of wild animals. Scientific Reports, 9(1), 1–10. **https://doi.org/10.1038/s41598-019-47684-6.**

508 Sapolsky, R. M. (2005). The influence of social hierarchy on primate health. Science (New York, N.Y.), 308(5722), 648–652. **https://doi.org/10.1126/science.1106477.**

509 King, B. J. (2014). How Animals Grieve (Reprint edition). University of Chicago Press.

510 Pinker, S. (2011). The Better Angels of Our Nature: Why Violence Has Declined. Penguin Books.

Rosling, H., Rönnlund, A. R., & Rosling, O. (2018). Factfulness: Ten Reasons We're Wrong About the World--and Why Things Are Better Than You Think. Flatiron Books.

511 Bartlett, T. (2019). Why Do People Love to Hate Steven Pinker? The Chronicle of Higher Education. **https://www.chronicle.com/article/why-do-people-love-to-hate-steven-pinker/** (Accessed August 6, 2020).

512 Park, A. (2012). George W. Bush and the Stem Cell Research Funding Ban. Time. **https://healthland.time.com/2012/08/21/legitimate-rape-todd-akin-and-other-politicians-who-confuse-science/slide/bush-bans-stem-cell-research/** (Accessed July 19, 2020).

513 Takahashi, K., & Yamanaka, S. (2006). Induction of pluripotent stem cells from mouse embryonic and adult fibroblast cultures by defined factors. Cell, 126(4), 663–676. **https://doi.org/10.1016/j.cell.2006.07.024.**

514 Narsinh, K. H., Plews, J., & Wu, J. C. (2011). Comparison of human induced pluripotent and embryonic stem cells: Fraternal or identical twins? Molecular Therapy: The Journal of the American Society of Gene Therapy, 19(4), 635–638. **https://doi.org/10.1038/mt.2011.41.**

515 Greenwood, J. (2019). Evolving Households: The Imprint of Technology on Life. The MIT Press.

516 Dubner, Stephen. (2016). The True Story of the Gender Pay Gap (Ep. 232). Freakonomics. **https://freakonomics.com/podcast/the-true-story-of-the-gender-pay-gap-a-new-freakonomics-radio-podcast/** (Accessed September 6, 2020).

CHAPTER 12

517 Hidden Brain: America's Changing Attitudes Toward Gay People. (2019). NPR.org. **https://www.npr.org/2019/04/17/714212984/hidden-brain-americas-changing-attitudes-toward-gay-people** (Accessed July 12, 2020).

518 Newport, F. (2019, October 29). Millennials' Religiosity Amidst the Rise of the Nones. Gallup.Com. **https://news.gallup.com/opinion/polling-matters/267920/millennials-religiosity-amidst-rise-nones.aspx** (Accessed April 16, 2021).

519 Gladwell, M. (2006). The Tipping Point: How Little Things Can Make a Big Difference (1st edition). Little, Brown and Company.

520 Lipka, M. (2019). 10 Facts about Atheists. Pew Research Center. **https://www.pewresearch.org/fact-tank/2019/12/06/10-facts-about-atheists/** (Accessed August 6, 2020).

521 Ehrman, B. D. (2009). Jesus, Interrupted: Revealing the Hidden Contradictions in the Bible (Reprint edition). HarperOne.

Qur'anic Contradictions - RationalWiki. (2020). **https://rationalwiki.org/wiki/Qur%27anic_contradictions** (Accessed August 6, 2020).

522 Petrosino, A., Petrosino, C. T., & Buehler, J. (2006). Scared straight and other juvenile awareness programs. Preventing Crime: What Works for Children, Offenders, Victims, and Places, 87–101. **https://doi.org/10.1007/1-4020-4244-2_6.**

Dubner, S. (2017). When Helping Hurts (Ep. 295). **https://freakonomics.com/podcast/when-helping-hurts/** (Accessed August 6, 2020).

523 Build Software Better, Together. GitHub. **https://github.com** (Accessed August 6, 2020).

524 https://aftermeatbook.com/contact/

525 I don't remember the exact date here; however, I remember that it was a 1 to 1.5 hour window of a particular day.

526 Coral Reproduction. Australian Government – Great Barrier Reef Marine Park Authority. **http://www.gbrmpa.gov.au/the-reef/corals/coral-reproduction** (Accessed December 26, 2019).

527 Link, H., Fuhrer, T., Gerosa, L., Zamboni, N., & Sauer, U. (2015). Real-time metabolome profiling of the metabolic switch between starvation and growth. Nature Methods, 12(11), 1091–1097. **https://doi.org/10.1038/nmeth.3584.**

528 Sekar, K., Rusconi, R., Sauls, J. T., Fuhrer, T., Noor, E., Nguyen, J., … Sauer, U. (2018). Synthesis and degradation of FtsZ quantitatively predict the first cell division in starved bacteria. Molecular Systems Biology, 14(11), 8623. **https://doi.org/10.15252/msb.20188623.**

529 Rao, J. (2016). Want to See the 2017 Solar Eclipse? Better Book Your Hotel Room Now. Space.com. **https://www.space.com/34545-book-reservations-for-great-american-eclipse.html** (Accessed August 20, 2020).

530 Navier–Stokes Equation. Clay Mathematics Institute. **http://www.claymath.org/millennium-problems/navier%E2%80%93stokes-equation** (Accessed December 26, 2019).

531 Haby, J. Vorticity Basics. National Weather Service **https://www.weather.gov/source/zhu/ZHU_Training_Page/Miscellaneous/vorticity/vorticity.html#VORT2** (Accessed December 26, 2019).

532 Pyrimidine Dimer. (2019). Wikipedia. **https://en.wikipedia.org/w/index.php?title=Pyrimidine_dimer&oldid=931253901** (Accessed December 26, 2019).

533 Somers, J. (2018). The Friendship That Made Google Huge. The New Yorker. **https://www.newyorker.com/magazine/2018/12/10/the-friendship-that-made-google-huge** (Accessed December 26, 2019).

534 Deutsch, D. (1998). The Fabric of Reality: The Science of Parallel Universes--and Its Implications. Penguin Books.

535 Penrose, R. (1996). Shadows of the Mind: A Search for the Missing Science of Consciousness (Reprint edition). Oxford University Press.

536 Hulbert, M. (2017). This Is How Many Fund Managers Actually Beat Index Funds. MarketWatch. **https://www.marketwatch.com/story/why-way-fewer-actively-managed-funds-beat-the-sp-than-we-thought-2017-04-24** (Accessed December 26, 2019).

537 Royal, J & O'Shea, A. (2018). What Is the Average Stock Market Return? NerdWallet. **https://www.nerdwallet.com/blog/investing/average-stock-market-return/** (Accessed December 26, 2019).

538 Harper, D. R. (2019). Hedge Funds: Higher Returns Or Just High Fees? Investopedia. **https://www.investopedia.com/articles/03/121003.asp** (December 26, 2019).

539 McAfee, A. (2019). More from Less: The Surprising Story of How We Learned to Prosper Using Fewer Resources—and What Happens Next (Illustrated edition). Scribner.

540 Taleb, N. N. (2010). The Black Swan: Second Edition: The Impact of the Highly Improbable: With a new section: "On Robustness and Fragility" (2nd ed. edition). Random House Trade Paperbacks.

541 Harari, Y. N. (2017). Homo Deus: A Brief History of Tomorrow (Illustrated edition). Harper.

542 Smil, V. (1999). Detonator of the population explosion. Nature, 400(6743), 415–415. **https://doi.org/10.1038/22672**.

543 Historical GDP of China. (2020). Wikipedia. **https://en.wikipedia.org/w/index.php?title=Historical_GDP_of_China&oldid=962431852** (Accessed August 19, 2020).

544 Kroeber, A. R. (2016). China's Economy: What Everyone Needs to Know® (1st edition). Oxford University Press.

545 Hanson, V. D. (2020). China Boomeranging. National Review. **https://www.nationalreview.com/2020/03/coronavirus-china-response-will-weaken-it-on-world-stage/** (Accessed August 19, 2020).

546 Graham-Harrison, E. & Kuo, L. (2020). China's Coronavirus Lockdown Strategy: Brutal but Effective | World News | The Guardian. **https://www.theguardian.com/world/2020/mar/19/chinas-coronavirus-lockdown-strategy-brutal-but-effective** (Accessed August 19, 2020).

547 COVID-19 Pandemic in the United States. (2020). Wikipedia. **https://en.wikipedia.org/w/index.php?title=COVID-19_pandemic_in_the_United_States&oldid=973915325** (Accessed August 19, 2020).

548 Holtz, D., Zhao, M., Benzell, S. G., Cao, C. Y., Rahimian, M. A., Yang, J., ... Aral, S. (2020). Interdependence and the cost of uncoordinated responses to COVID-19. Proceedings of the National Academy of Sciences, 202009522. **https://doi.org/10.1073/pnas.2009522117**.

549 Chen, D. (2015). How China Holds Politicians Accountable. TODAYonline. **https://www.todayonline.com/chinaindia/china/how-china-holds-politicians-accountable** (Accessed December 26, 2019).

550 Xu, H., Wu S., & Zheng, E. Bribery & Corruption 2021 | China. GLI - Global Legal Insights. **https://www.globallegalinsights.com/practice-areas/bribery-and-corruption-laws-and-regulations/china** (Accessed August 19, 2020).

551 Worland, J. (2017). It Didn't Take Long for China to Fill America's Shoes on Climate Change. Time. **https://time.com/4810846/china-energy-climate-change-paris-agreement/** (Accessed August 19, 2020).

552 Ruck, D. J., Bentley, R. A., & Lawson, D. J. (2018). Religious change preceded economic change in the 20th century. Science Advances, 4(7), 1–8. **https://doi.org/10.1126/sciadv.aar8680.**

553 Smil, V. (2005). Creating the Twentieth Century: Technical Innovations of 1867-1914 and Their Lasting Impact (Illustrated edition). Oxford University Press.